CW01613206

ELECTRIC FLIGHT

ELECTRIC FLIGHT

by Dave Day

ARGUS BOOKS LIMITED

ARGUS BOOKS LIMITED
Model & Allied Publications,
Wolsey House, Wolsey Road,
Hemel Hempstead, Herts, HP2 4SS

© Argus Books Ltd

First published 1983
ISBN 0 85242 822 7.

All rights reserved. No part of this publication may be reproduced in any form or by any person without permission of the publisher.

Phototypesetting by M & G Studios Ltd, London NW 10
Repro by Tenreck Ltd, London NW 10

Printed & bound in Great Britain By R.J. Acford
Chichester, Sussex

Contents

ACKNOWLEDGEMENTS

Additional material and research for this book have been contributed by

Dave Dyer
Peter Holland
Vic Smeed

Thanks are also due to manufacturers and individuals whose assistance with illustrations is very much appreciated.

Introduction

If we are to believe the records, an electric powered model aircraft has been a feasible possibility since the year 1909.

However, it was not until the late 1950s that electric free flight models became reasonably practical, while ultra-light R/C models followed in the mid 1960s. Today one can purchase a complete almost-ready-to-fly (ARTF) electric powered R/C model from leading toy shops and a large range of motors, batteries and complete power packs are available which can power any type or size of electric aircraft, be it F/F, C/L, or R/C.

In a pollution-conscious, noise-hating world, the air should, logically, be full of electric powered model aircraft. Needless to say, this is not so, though it is not easy to determine just why. In fact, electric powered models are still very much in the minority.

The Europeans have regular contests for this type of model and have advanced to the point where the survival of heavily loaded, high performance models is a considerable tribute to their piloting ability and the reliability of modern R/C equipment.

It is not the writer's intention to promote the development of this type of aircraft but rather to illustrate the ease and practicability of the modern electric powered model aircraft.

Dave Day

Photo 1. The interior of Col. Taplin's Radio Queen. *The forward bay holds a hard-valve receiver, suspended on its side by rubber bands, with HT & LT batteries beneath. Next are the banks of silver-zinc cells with, behind them, an E.D. clockwork timer mounted above a three speed controller, the timer providing an overall shut-off if the radio link failed. Just in the bottom of the picture is the clockwork escapement, connected to the rudder by two external cords, which was the only other control.*

Photo 2. The 30oz 24v government surplus Emerson motor which on 30v and 8 amps would lift the 7 ft, 8lb model off a smooth runway. The Model was the seven year old Radio Queen *prototype, normally flown with a 3½cc diesel.*

Chapter One

History

In an issue of 'Model Engineer' dated October 21st 1909, there appeared an account of an electric powered free flight (F/F) model which, it was claimed, had been constructed and flown. The validity of this claim was questioned some 48 years later by the renowned Colonel H.J. Taplin, who on the 30th June 1957 made the first officially recorded electric powered radio controlled model flight. Col. Taplin was prompted to offer £100 to anyone who could produce a replica of the 1909 model and make it fly.

Electrical power for the Col.'s model was supplied by 25 silver/zinc cells weighing a total of a little over 28oz. Motor weight was 30oz. and the total model weight 8lbs.

Further developments in the field came from the great German pioneer, Fred Militky, who first achieved a successful flight with a F/F model in October 1957 using a special motor.

In February 1959, a Dr. Ing. Fritz Faulhaber walked into the offices of the German magazine 'Modell' and enquired if a motor he had developed for use in remote controlled camera shutters might be of use in modelling. This was the now famous Micromax, a coreless motor with integral gearbox.

With this motor, Militky developed a series of designs for the *Graupner* firm which brought practical electric flying models within the reach of the average modeller. (See fig. 2).

At this time, the nickel/cadmium cell was still to appear and batteries used for these early flights were either small sealed lead/acid batteries (accumu-

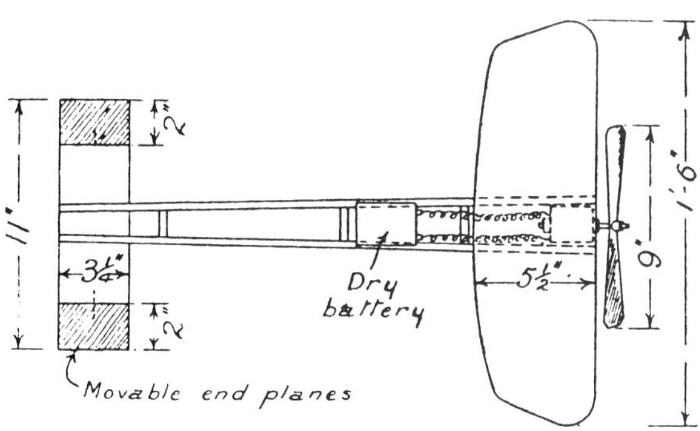

DIAGRAMMATIC PLAN OF AEROPLANE.

Fig. 1. Published in "Model Engineer' in 1909, this 18 in. span model with a "heavy cardboard" wing and 9 in. x 1½ oak propeller, weighing nearly 2 lbs, was claimed to have flown for eight minutes. The builder also claimed an o.o.s flight with a 12 sq.ft model carrying a 6 lb accumulator...

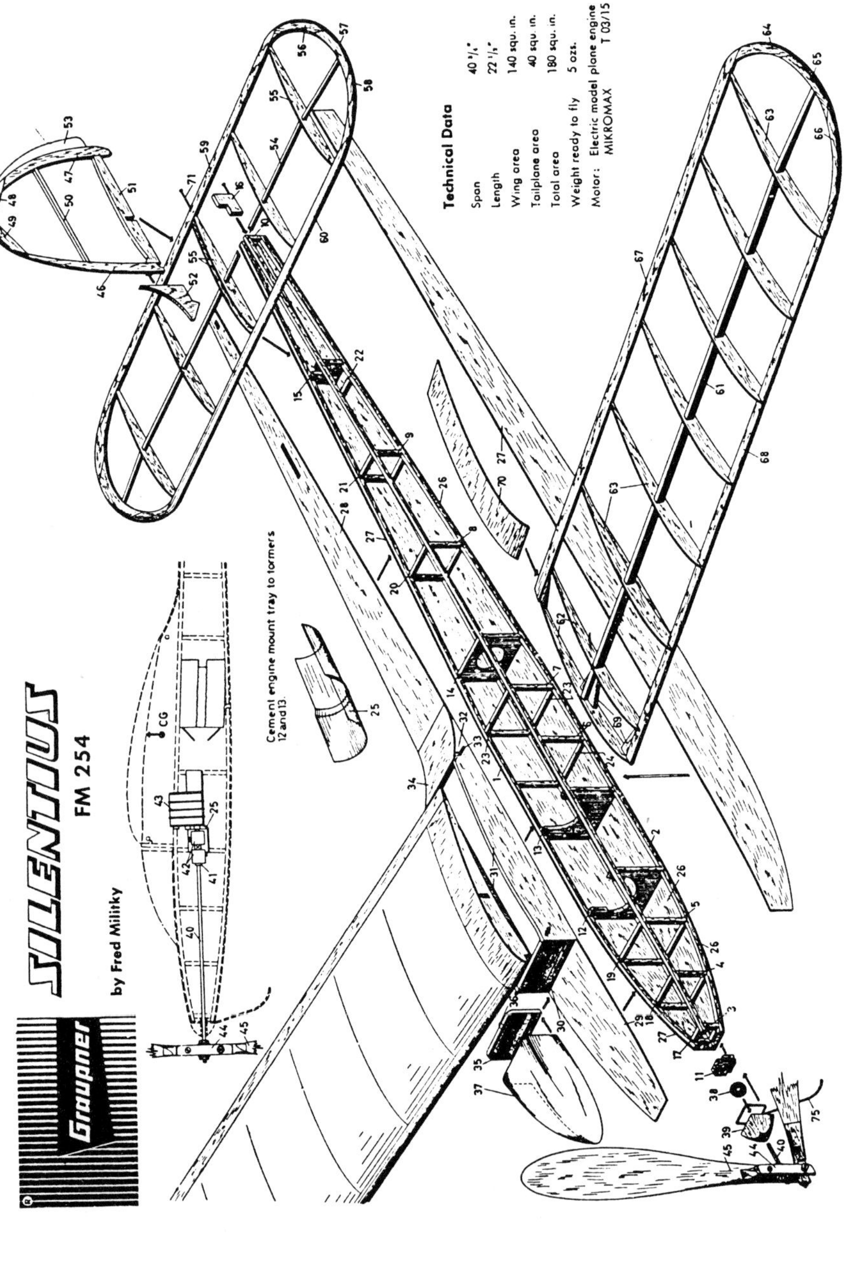

FIG. 2 SILENTIUS BY FRED MILITKY

MICROMAX TO3 TESTS

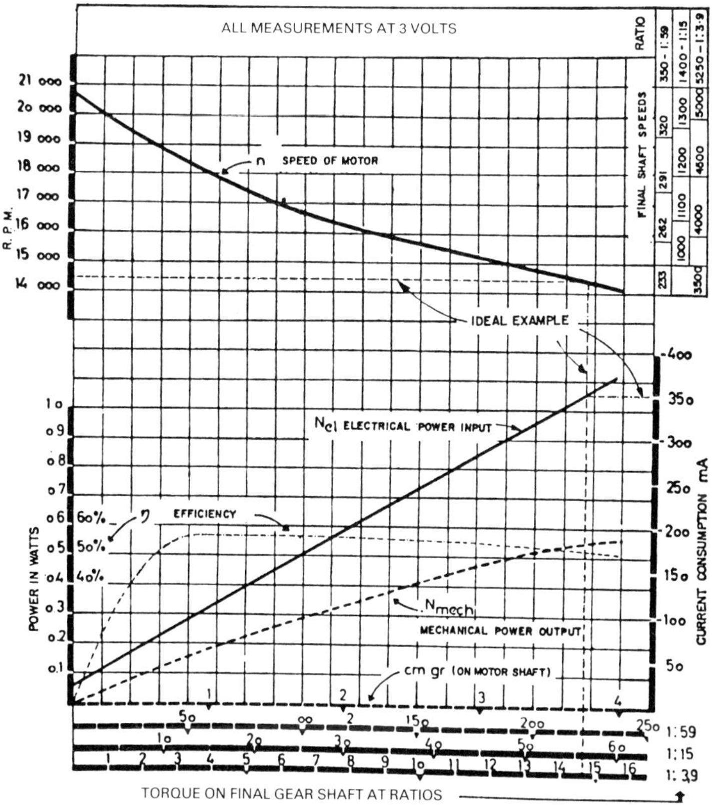

FIG. 3

lators) or salt water activated cells.

Militky went on to develop lightweight R/C models, then the more sophisticated *Graupner* 'Mosquito' and 'Hi-Fly' designs and was to play a leading part in the design of a man-carrying electric aircraft.

Following a lull in the commercial side of electric flight, 1972 saw the introduction of the *Mattel* Super Star. This ready-formed plastic and foam aircraft was powered by a Japanese *Mabuchi* motor taking its current from two fast charge nickel/cadmium cells. A novel feature was the inclusion of a cam, gear driven from the motor, which moved the rudder and gave a degree of flight programming. Recharging was accomplished by connecting the model to a 6 volt zinc/carbon 'lantern' battery. (See photos 4 and 5).

This principle was continued in the *Mabuchi* A.1. power unit which appeared in 1974. Here provision was made for the modeller to adapt things to his own requirements, with the 6 volt

charger being fitted with 4 – U2 ('D' size) batteries and the nickel/cadmium cells being removed from the power unit for charging. Unlike the Mattel unit, however, the propeller was driven directly rather than via reduction gearing. This unit is still available. (See photo. 8). At this point, numerous units began to ap-

Photo 3. The late Fred Militky with his Hi-Fly*, the starting point for the first manned electric plane, the* MBE 1.

Photo 4. The Mattel Super Star *appeared in 1972, a plastic and foam free-flight model with cam programming.*

pear from Japan, Germany and the USA. These catered for almost any type of model from small F/F aircraft to large R/C models.

Photo 5. Gear reduction was a feature of the Mattel power unit shown above.

Photos 6 and 7. Astro-Flight motors on a display stand and, right, a simple power pylon for a self-launching sailplane.

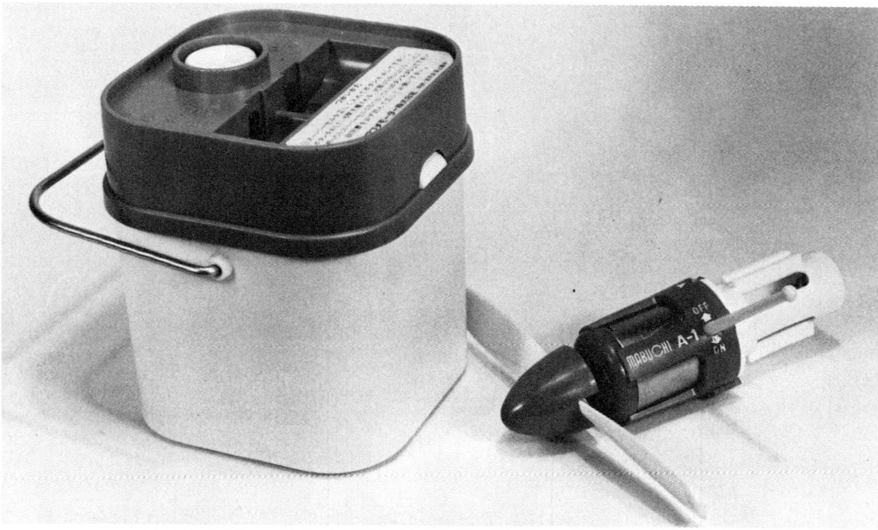

While development continued in these countries, Britain appeared to lose interest. The advent of almost-ready-to-fly models such as the *Nitto* 'Kitty', *Kyosho* 'Sportavia' and *Acoms* 'Cessna' did little to improve the situation (In fact the 'Sportavia' is still unknown in the U.K., while only small numbers of the 'Kitty' were seen) and only a small band of enthusiasts still persevere.

In Europe, electric R/C models have gone from strength to strength with numerous contests being held for duration, aerobatics, pylon racing, scale etc. Motors using samarium cobalt magnets and fed from 30 or more cells have produced performance levels nearly equal to models powered by internal combustion engines.

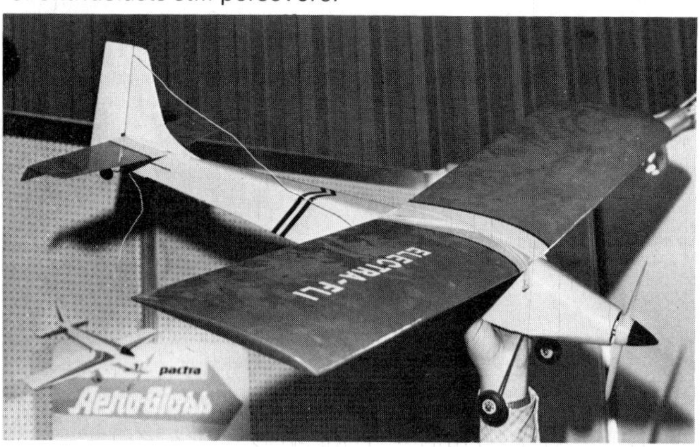

Photo 8 (above). The Mabuchi A1 power unit, first seen in 1974, with its charger, using dry batteries to charge the flight nicads.

Photo 9. One of the early radio control models commercially available for electric flight was the Electra-Fli.

Chapter Two

The Model

Possibly due to the fact that early electric models were of necessity built as lightly as possible, there is a convention that any electric powered model *must* be lightly built. This approach has an obvious disadvantage in that a heavy landing will inevitably cause damage due to the structure's inability to retain the heavy batteries.

Since it is certain that any *practical* electric powered model will have a relatively heavy wing loading, it is illogical

Photo 10. Shown in flight is the Cotswold Electric Tracer, *a Club 20 pylon-racing model in electric form. Installation in the GRP fuselage is shown below.*

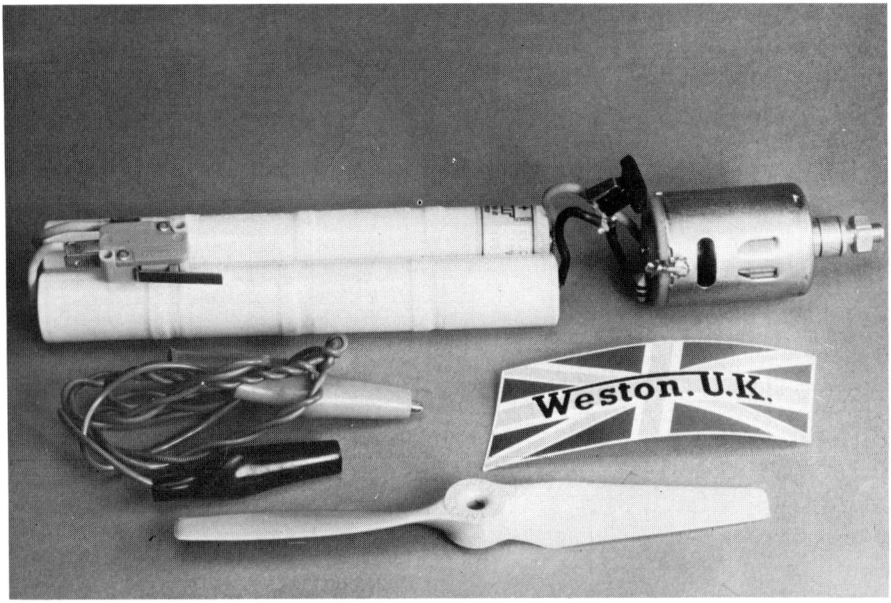
Weston.U.K.

FIG 4. Method of obtaining aileron differential.

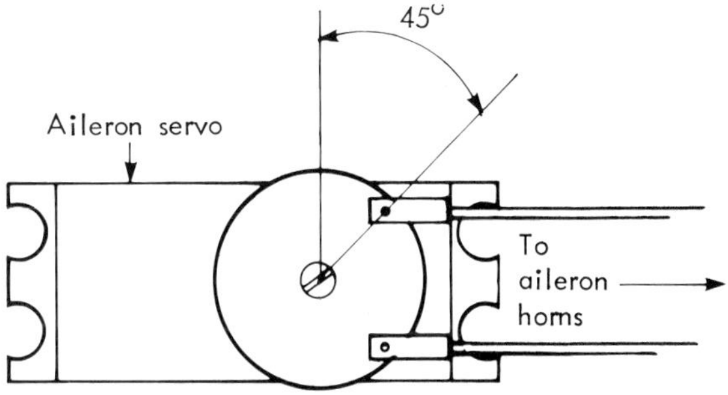

to take extreme lightening measures with the structure which can save no more than a few ounces.

A MORE PRACTICAL WAY

Experiments by the writer and others have shown that a standard 'Club 20' airframe with GRP (glass reinforced plastic) fuselage and veneered foam wings can be converted to a practical electric model of more than adequate performance using inexpensive power units (see photo. 10).

Two Club 20 kit manufacturers have produced special versions of their models, these being the *Cotswold* 'Electric Tracer' and *Flair* 'Darlec'. A special power unit was produced by Weston (UK) Ltd. for these aircraft consisting of *Mabuchi* RS.540 motor with eight 1.2AH *Saft* cells (see photo. 11).

A GRP fuselage has certain advantages over other types of construction in that it combines strength with flexibility and is able to absorb a considerable impact without damage. This is especially true where no undercarriage is fitted since there is only a thin layer of GRP between the batteries and the ground and very little of the landing shock is transmitted through the airframes.

Opposite, top photo 11. Weston (UK) Ltd. market this flight pack consisting of a Mabuchi RS.540 and eight 1.2 a/h Saft cells.

Opposite, bottom, photo 12. Nitto Kitty *battery holder is mounted on fuselage underside and can break clear without major damage.*

OTHER STRUCTURES

Where other forms of construction are used, it is worth studying certain ARTF models (see Chapter 6) where the batteries are located in recesses in the fuselage bottom and arranged to break free on impact. In effect, the model is attached to the battery, rather than the battery to the model (see photo. 12).

SCALE MODELS

For scale aircraft, there is more of a problem. Most existing designs for I.C. engines feature ply doublers in the forward fuselage area which enclose the tank bay. This bay is usually just above the undercarriage and forms an ideal site for battery stowage.

One such conversion was performed on the RCM & E plans 'Fokker D.VII' design by David Boddington. Power was provided by an MFA 'Olympus' geared unit fitted with a Mabuchi RS 540 motor and driving an 11in × 8in propeller. This

Photo 13. Nose of Fokker DVII model designed for .19 (3.2cc) engines but flying with an MFA Olympus *unit with eight cells. (Also on cover)*

FIG. 5 Different props on same motor/battery combination

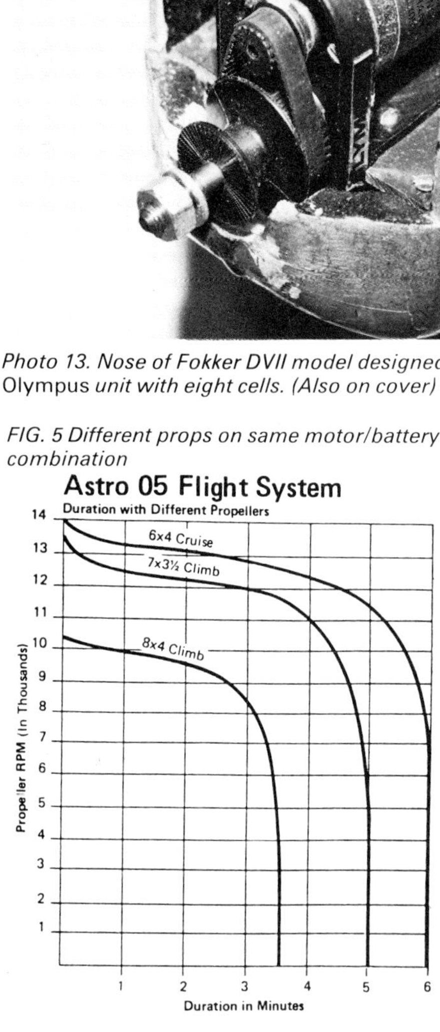

was fitted to a paxolin plate which was, in turn, bolted to the existing engine bearers. Current was supplied by eight 1.2AH nickel/cadmium cells installed in the tank compartment (see photo. 13).

CONTROL PROBLEMS

One problem associated with conversions of existing designs, like the Fokker DVII above, is a lack of aileron response, particularly during the initial climb. This is due to initial air speed being marginal until sufficient height is reached to enable the nose to be lowered and speed to be built up.

Compared to an equivalent I.C. powered model, an increase in aileron area or movement of roughly 50% is advisable, coupled with a degree of aileron differential (see fig. 4).

LANDING

Landing speeds will normally be high and it is advisable to concentrate on making the *smoothest* possible landing rather than the *slowest*. It follows from this that the more experienced, capable flyer is more likely to be successful in the flying of electric powered models.

CHOICE OF PROPELLER

Probably the most important item in the success or failure of any electric model, the propeller has to satisfy many requirements and can only be, at best, a compromise.

A propeller which is too large will overload the motor and cause it to draw

Photo 14. Typical self-launching electric powered glider, with folding pusher airscrew. This particular one no longer available.

excessive current, giving short flights and possible motor damage due to overheating.

Too small a propeller will give excellent duration but will result in poor model performance.

The type of motor winding and the number of cells in the battery will also influence the propeller size. Note also that, for a given motor, increasing the number of cells in the battery will normally require a *smaller* propeller to avoid overloading.

With geared units, surprisingly large propellers are needed to give adequate performance. A rough rule is to go on

Photo 15. Free-flight electric competition duration models tend to high aspect ratio wings and slim fuselages and are, effectively, powered gliders.

increasing the size until there is a marked drop in the duration of motor run. Since the propeller RPM will be relatively low, rather more pitch will be needed than would normally be expected. The Fokker DVII already mentioned needed 11in diameter × 8in pitch before it could be persuaded to fly.

Where a commercial unit is used, there will normally be a propeller size recommended by the manufacturer which will give a useful starting point (see fig. 5). The following table gives recommendations for the well known Astro-Flight series of motor/battery units and is a useful guide for other applications.

RECOMMENDED APPLICATIONS OF VARIOUS ASTRO-FLIGHT SYSTEMS

Motor System	Type of Model	Prop dia x pitch (in)	Flying Wt. (oz)	Wing Area (sq. in.)
020FF	FF	5¼ x 3	7–8	150–220
020 FF	FF Scale	5⅛ x 3, 6 x 3	8–12	165–250
020 R/C	R/C 1 ch.	5¼ x 3	13–17	165–250
05	Sailplane	7 x 3½	36–44	500–750
05	Sport	6 x 4	32–38	280–390
05	Scale	7 x 3½	34–40	300–420
075 (12)	Sport	7 x 4	40–60	330–500
075 (12)	Sailplane	8 x 4	42–50	750–1000
075 (8)	Sport/Scale	8 x 5	36–42	320–450
10	Sport/Scale	7 x 4	45–55	350–550
10G (16)	Scale	13 x 8	60–72	600–1000
15	Sport	7 x 4	55–65	400–550
15	Scale	8 x 4	55–65	400–600
15G	Sailplane,	14 x 7, Old Timer 16 x 7	50–60	700–1000
15G (16)	Scale	14 x 8, 13 x 8	60–72	700–1000
25	Sport/Scale	9 x 6	75–85	600–750
25G	Sailplane, Old Timer	18 x 8	96–112	1000–1500
25G	Scale	16 x 12, 18 x 10	112–128	850–1500
02 Twin	Scale	5¼ x 3	28–32	350–450
05 Twin	Scale	6 x 4	60–70	475–600
15 Twin	Scale	7 x 4	96–126	700–900
25 Twin	Scale	8 x 6	130–144	750–950

Chapter Three

Motors

MABUCHI

The Mabuchi motor is by far the most common make in use and for our purposes comes in three basic sizes, the RS. 380, RS. 540 and RS. 550. There has been a considerable amount of development carried out on the RS. 540 by manufacturers of electric flight equipment and even more work done by the electric R/C car world.

As a result, there are a staggering number of different types of this motor and, to a lesser extent, the RS. 550. These differences are due to the different sizes of copper wire used to wind the armature, and the number of turns applied.

When using a motor supplied by a notable electric flight specialist, no problems should arise, but any other motor should be treated with suspicion. Most of the 'hot' electric car motors are unsuitable for our purpose, while others *can* be used if propeller size and battery voltage are restricted. One electric car motor which *can* be used for electric flight is the so-called 'standard' RS. 540 having 35 turns of 23 swg wire on each pole of the armature. This is roughly equivalent to the *Astro-Flight* 05, *MFA* Hummingbird 15, *Ripmax* Cyclone 15, etc. (See photo. 16).

If using an unknown motor of the RS. 540/550 type, don't be surprised if you get very little power and a very long running time since there are several versions of these motors with very high resistance armatures.

A motor intended for electric car use will normally carry some identification to that effect.

The RS. 380 motor is somewhat smaller than the RS. 540/550 and is used, with reduction gearing, in several ARTF models (Nitto 'Kitty', Acoms 'Cessna' etc.).

Photo 16
Possibly the most widely used basis for specialised motors for aircraft, boats and cars is the Mabuchi RS.540

Astro 05

Weight 150gms

Voltage at Motor Terminals 8.8V

Prop	Current(A)	R.P.M.
6 x 3 Cox	6.0	13,000
6 x 3 Tornado	6.45	12,900
6 x 4 Cox	9.3	11,700
6 x 4 Topflite	7.2	12,400
7 x 4 Topflite	11.25	11,800

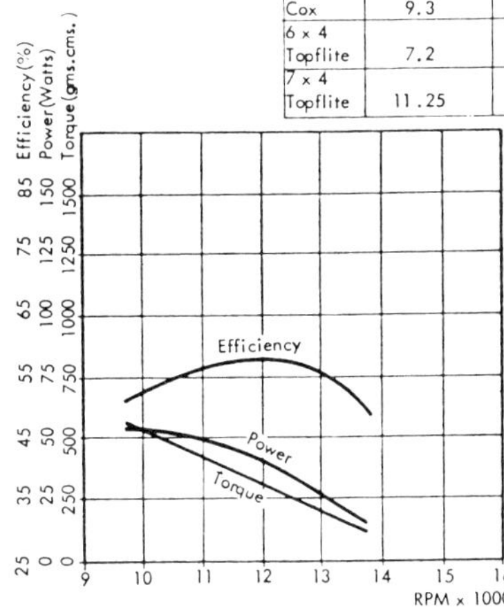

FIG. 6 Astro 05 motor performance

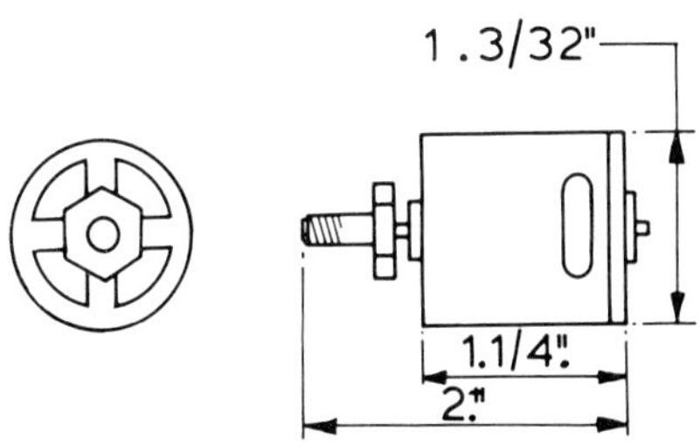

ASTRO 02

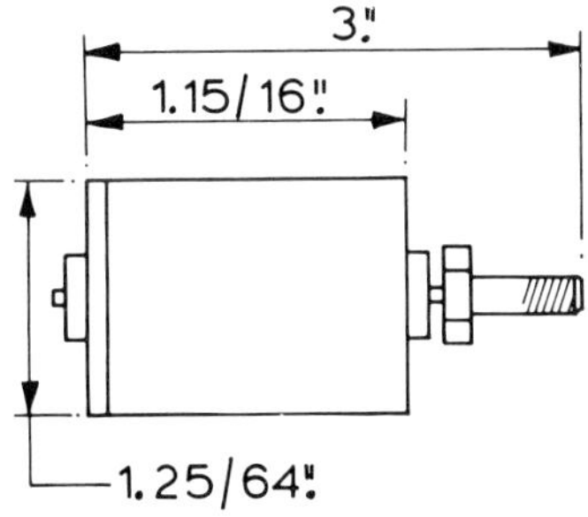

ASTRO 05

ASTRO-FLIGHT 02

Intended for F/F or lightweight R/C this motor has a weight of 50gm and is for use with a 4 cell (4.8 volt) battery. Recommended prop size is 5½in x 3in.

ASTRO-FLIGHT 05

Most popular motor size. Suitable for sport or scale R/C and powered gliders. Use with 8 cell battery (9.6 volts).
Prop sizes from 6in x 3in to 8in x 4in.
Weight 155gm (See fig. 6).

Astro 10

Voltage at Motor Terminals 13.2V

Prop	Current(A)	R.P.M.
6 x 4 Cox	5.4	11,400
7 x 4 Topflite	6.9	10,900
7 x 6 Taipan	8.7	10,100

Weight 255gms

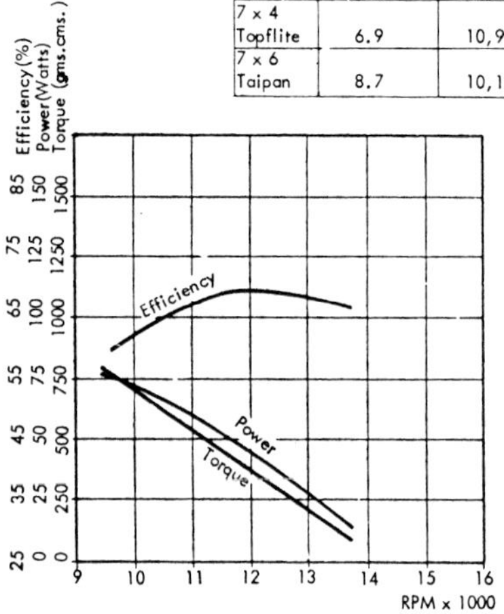

FIG. 7 Astro 10 motor performance

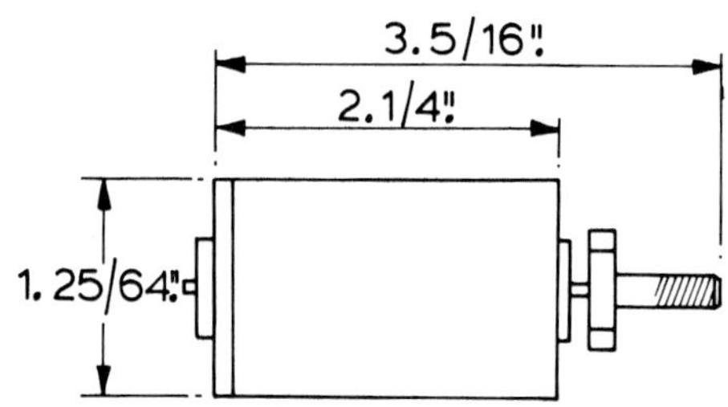

ASTRO 075 and 10

ASTRO-FLIGHT 075

Slightly longer version of 05 motor. Has more torque and can be used with up to 12 cells (14.4 volts).
Weight 160gm. Prop sizes from 7in x 4in to 9in x 4in.

ASTRO-FLIGHT 10

This motor can be used with up to 16 cells for which the recommended prop is 7in x 4in. Weight 255gm.
A geared version is available which permits the use of a 13in x 8in prop (See fig. 7).

Astro 15

Weight 245 gms

Voltage at Motor Terminals 17.6V

Prop	Current(A)	R.P.M.
6 x 4 Cox	4.8	12,200
7 x 4 Topflite	6.3	11,500
7 x 6 Taipan	7.8	10,500
8 x 4 Tornado	8.4	10,200

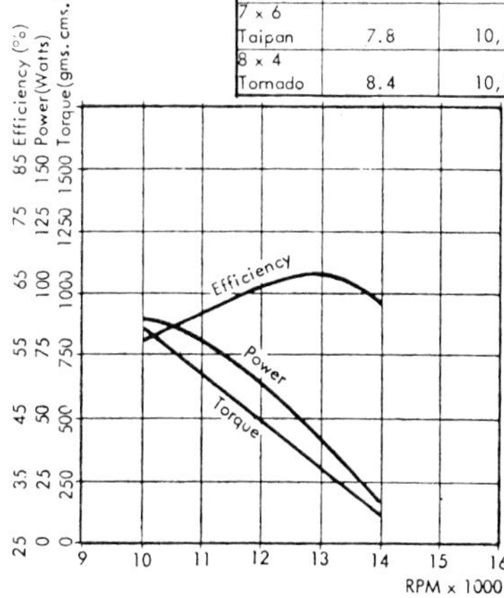

FIG. 8 Astro 15 motor performance

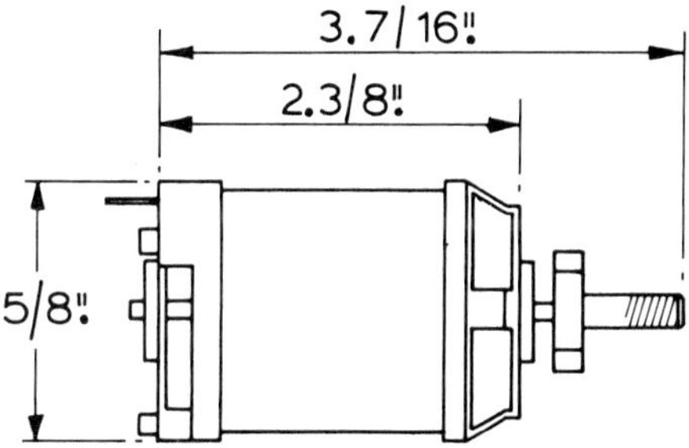

ASTRO 15

ASTRO-FLIGHT 15

Generally similar to the 10 motor but with a different armature winding allowing the use of higher voltage and larger propellers. Weight 255gm. Recommended prop with 16 cells 8in x 4in. Geared version will turn 16in x 7in. (See fig. 8).

Astro 25

Weight - 370 gms

Voltage at Motor Terminals 17.6V

Prop	Current(A)	R.P.M.
8 x 6 Topflite	10.5	9,900
9 x 4 Tornado	10.8	9,800
9 x 6 Kavan	15.6	8,400
10 x 4 Kavan	11.7	9,600

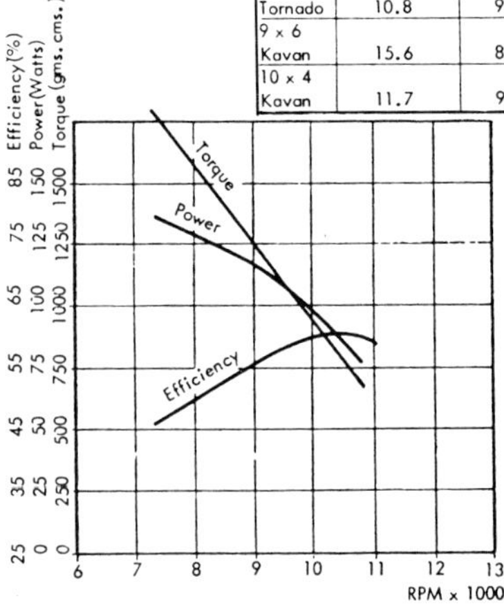

FIG. 9 Astro 25 motor performance

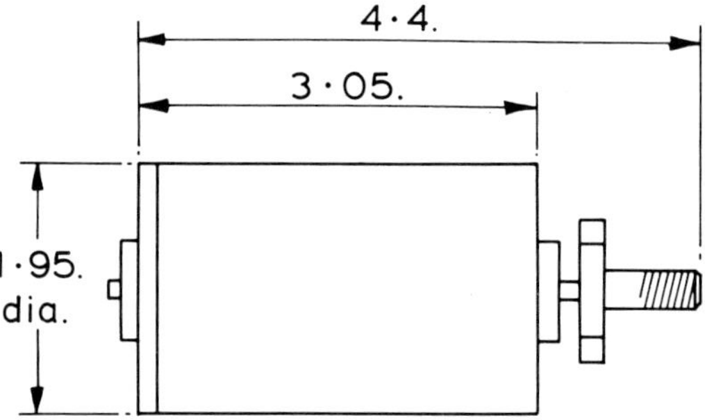

ASTRO 25

ASTRO-FLIGHT 25

On 16 cells this motor will turn a 9in x 6in prop, while the geared version can accommodate a 16in x 12in or 16in x 10in. Weight 370gm. (See fig. 9).

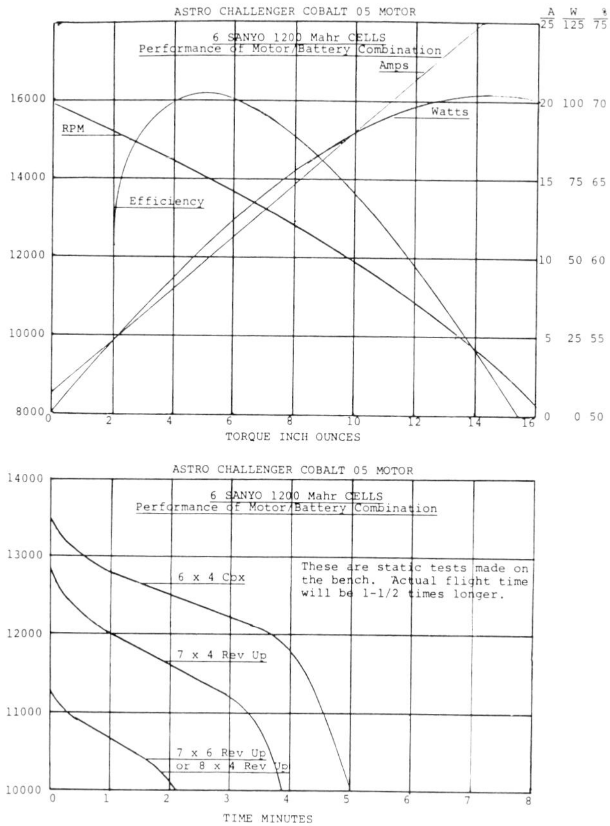

FIG. 10 Astro 05 Cobalt motor performance on 6 cells.

ASTRO-FLIGHT CHALLENGER COBALT 05

While weighing the same as the *Astro-Flight* 05, this motor is longer and of less diameter. It features large, heavy duty brushgear and a samarium cobalt magnet giving higher torque. Recommended propeller sizes range from 5½in x 4in to 8in x 4in when used with 6, 7 or 8 cells (See figs. 10,11 and 12). Weight 155gm.

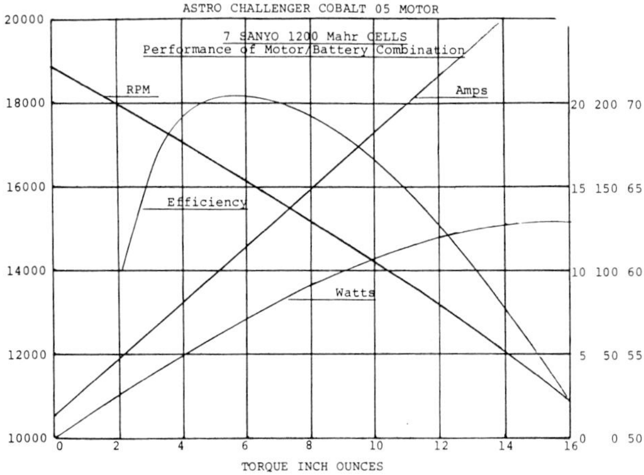

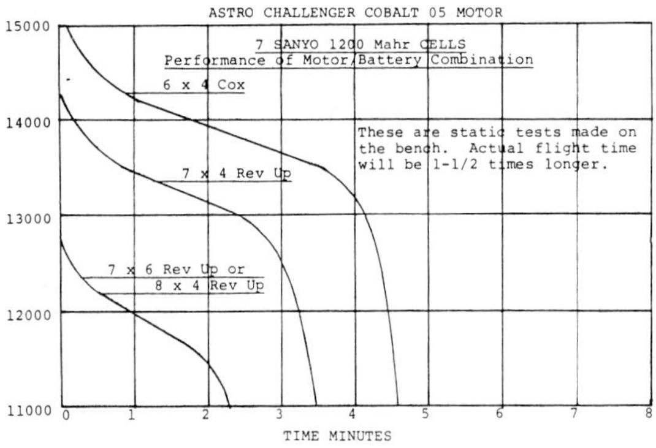

FIG. 11 Astro 05 Cobalt motor performance on 7 cells.

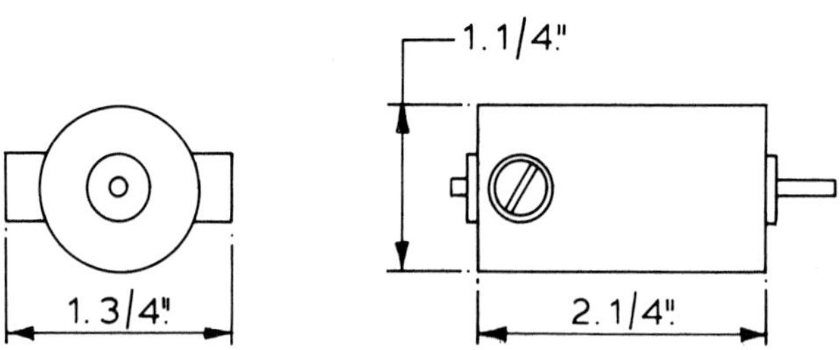

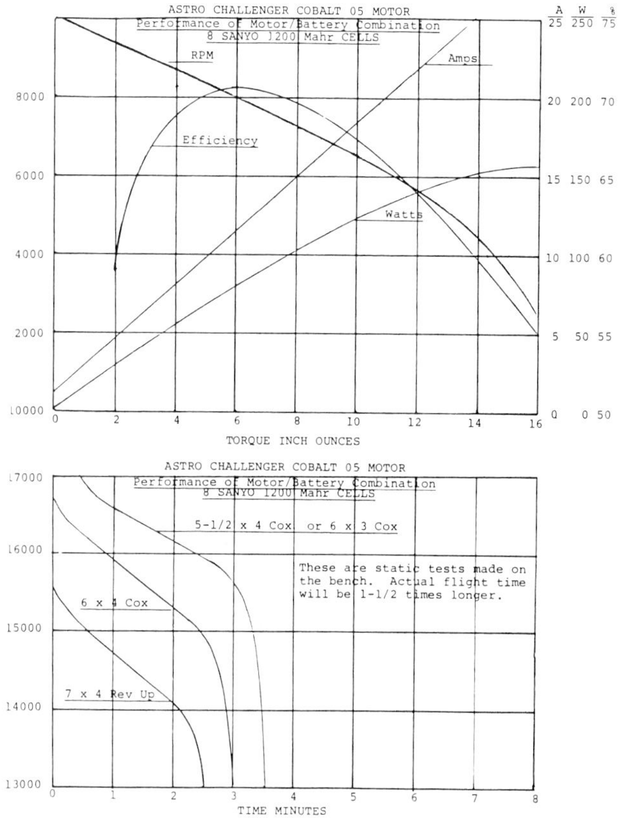

Fig. 12 Astro 05 Cobalt motor performance on 8 cells.

M.F.A. motors

Weight 155 gms

Voltage at Motor Terminals 8.8V

Prop	Current(A)	R.P.M.
6 × 3 Cox	8.4	14,600
6 × 3 Tornado	8.85	14,400
6 × 4 Cox	12.45	12,800
6 × 4 Topflite	10.2	13,900
7 × 4 Topflite	16.2	12,100
7 × 6 Taipan	20.25	11,000
8 × 4 Tornado	20.4	10,400

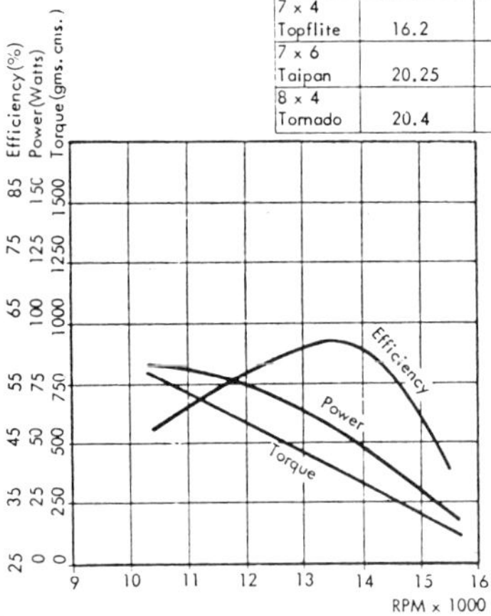

FIG. 13 MFA 15 motor performance.

Motor — Hummingbird '20'

Propeller	V_{in} (V)	I_{in} (A)	RPM	P_{out} (W)
Kavan 8 × 6	6	9.9	5830	31
	8	15.3	7330	62
	8.8	17.4	7900	73
	10	21.0	8550	92
Zinger (wood) 7 × 6	6	6.9	6860	26
	8	10.8	8550	53
	8.8	12.9	9780	68
	10	15.3	10620	90
Topflite 7 × 4	6	5.1	7330	23
	8	7.5	9310	36
	8.8	9.3	10340	48
	10	11.4	11370	70
	12	15.0	13070	101
Tornado 6 × 3	8	4.2	10670	16
	8.8	5.1	11560	29
	10	5.4	12130	37
	12	7.5	14950	58
	14	10.5	17390	94

Motor — Hummingbird '20' with Olympus

Propeller	V_{in} (V)	I_{in} (A)	RPM	P_{out} (W)
11 × 6 Kavan	6	4.8	3570	19
	8	7.5	4600	35
	10	10.2	5450	59
	12	13.2	6300	83
10 × 6 Kavan	6	3.9	3760	15
	8	5.4	4700	29
	10	7.5	5730	44
	12	9.6	6580	61
9 × 6 Topflite (wood)	6	3.9	3670	14
	8	5.4	4750	25
	10	7.8	5830	45
	12	9.6	6580	66

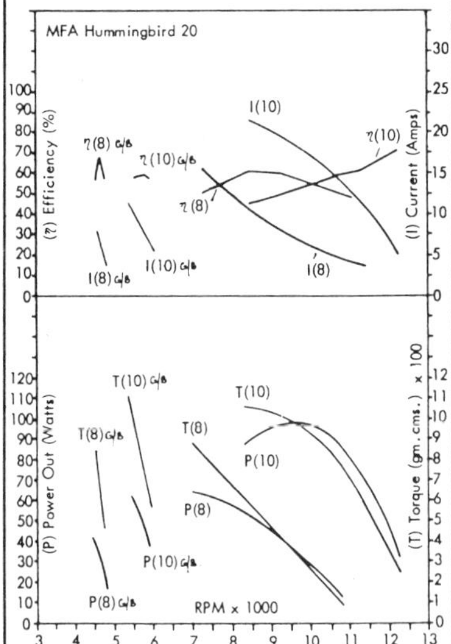

FIG. 14 MFA 15 motor performance.

MFA HUMMINGBIRD 02

Generally similar to Astro-Flight 02. Weight 65gm.

MFA HUMMINGBIRD 15

Generally similar to Astro-Flight 05. Weight 155gm. (See fig. 13).

MFA HUMMINGBIRD 20

Generally similar to Astro-Flight 075. Weight 160gm. (See fig. 14).

MFA HUMMINGBIRD 30

Generally similar to Astro-Flight 15. Weight 275gm.

Ripmax Cyclone 15

Weight 155 gms

Voltage at Motor Terminals 8.8V

Prop	Current(A)	R. P. M.
6 x 3 Cox	8.4	14,400
6 x 3 Tornado	9.3	14,600
6 x 4 Cox	13.5	13,200
6 x 4 Topflite	11.1	14,100
7 x 4 Topflite	17.25	12,300
7 x 6 Taipan	21.3	11,100
8 x 4 Tornado	23.1	10,700

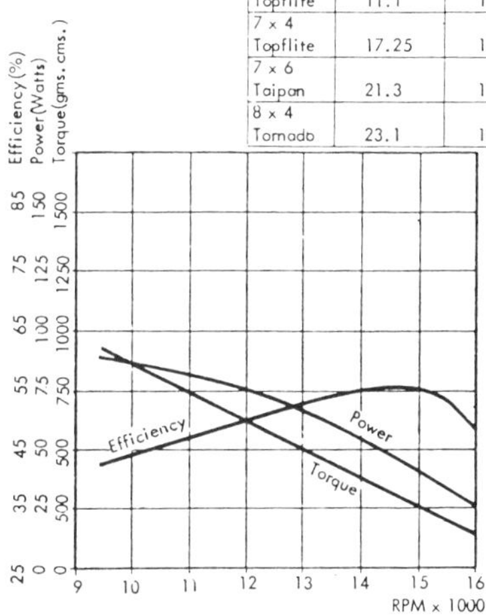

FIG. 15 Ripmax Cyclone 15 motor performance.

RIPMAX CYCLONE 15

Generally similar to Astro-Flight 05. Weight 155gm. (See fig. 15).

RIPMAX BULLET 30

While similar in size to the Astro-Flight 25 this motor is considerably more powerful. Weight 415gm. (See fig. 16).

Ripmax Bullet 30

Weight 415 gms

Voltage at Motor Terminals 17.6V

Prop	Current(A)	R. P. M.
8 x 6 Topflite	14.7	11,000
9 x 4 Tornado	15.3	10,900
9 x 6 Kavan	21.0	9,200
10 x 4 Kavan	16.5	10,600

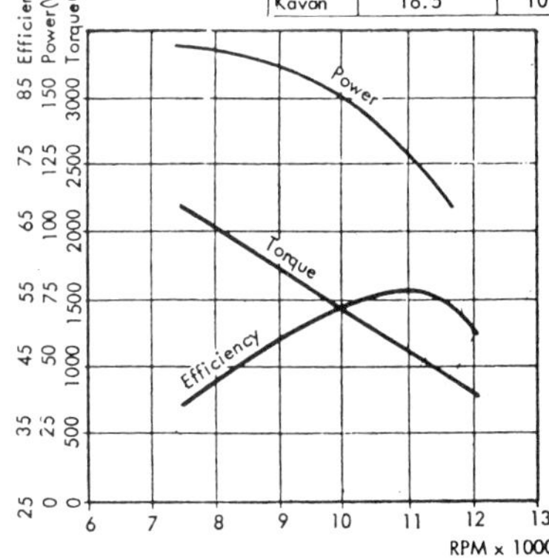

FIG. 16 Ripmax Bullet 30 motor performance.

Motor — Keller 30/8

Propeller	V_{in} (V)	I_{in} (A)	RPM	P_{out} (W)
Topflite (Wood) 9 × 6	6	11.7	5920	37
	8	19.8	7710	89
	10	30	9120	14.1
Kavan 8 ×6	6	9.9	6200	33
	8	16.5	7990	74
	10	26.7	9590	133
Zinger (Wood) 7 × 6	8	10.5	8650	47
	10	15.6	10810	92
	12	22.5	12310	152
Topflite 7 × 4	8	7.5	9120	28
	10	10.5	11280	61
	12	15.0	13440	93
	14	19.5	15510	143

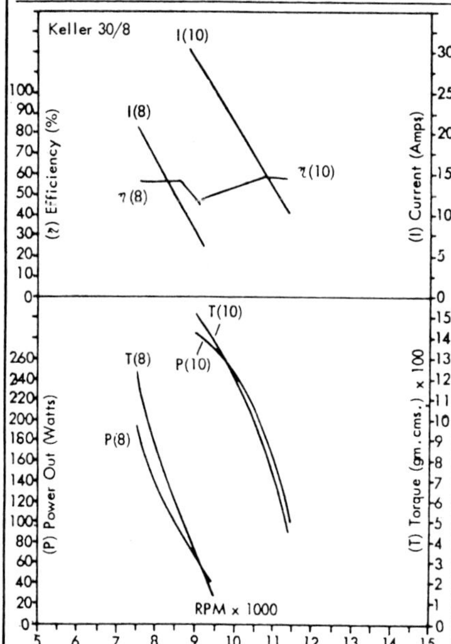

FIG. 17 Keller 30/8 motor performance.

KELLER 30/8

Intended for use with 8 cells, this motor has samarium cobalt magnets and weighs 250 gm. Prop sizes from 7in x 4in to 9in x 6in. (See fig. 17).

Motor — Keller 30/12

Propeller	V_{in} (V)	I_{in} (A)	RPM	P_{out} (W)
Kavan 11 × 6	8	12.9	5550	60
	10	19.5	6670	98
	12	30	7710	160
Kavan 10 × 6	8	9.6	6110	47
	10	13.8	6860	74
	12	19.2	8550	118
	14	26.4	9490	168
Topflite 9 × 6 (Wood)	8	8.7	6020	37
	10	12.9	6950	86
	12	18.0	8200	114
	14	24.6	9590	148
Kavan 8 × 6	8	7.5	6020	28
	10	10.5	7330	56
	12	14.4	8740	94
	14	19.2	9780	136
	16	29.4	10190	196
Zinger 7 × 6 (Wood)	10	7.2	8180	38
	12	9.3	9780	68
	14	12.6	11190	103
	16	16.5	13070	161
	18	19.5	14100	185
	20	28.6	14850	252
Topflite 7 × 4	10	6.3	9030	35
	12	8.1	11190	60
	14	9.3	11840	82
	16	11.4	13160	112
	18	13.2	14290	132
	20	18.0	16730	193

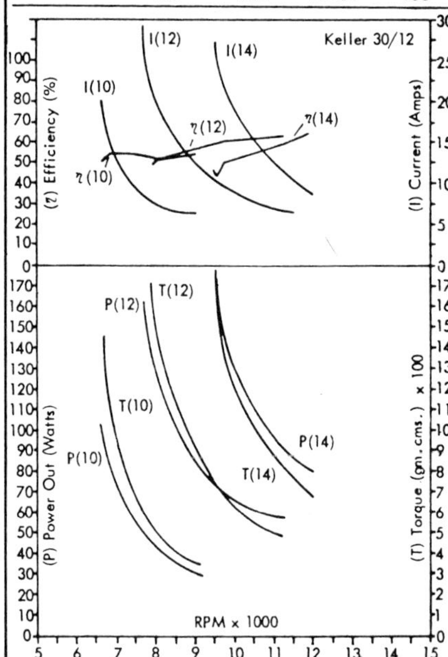

FIG. 18 Keller 30/12 motor performance.

Motor — Keller 50/10

Propeller	V_{in} (V)	I_{in} (A)	RPM	P_{out} (W)
Kavan 12 × 6	8	19.9	5922	106
	10	30.0	7240	195
Kavan 11 × 6	8	15.0	6300	83
	10	21.6	7520	145
	12	30.0	8840	245
Kavan 10 × 6	8	10.5	6390	51
	10	10	7900	104
	12	20.1	9310	165
	14	27.9	10900	269
Topflite (Wood) 9 × 6	8	10.2	6580	51
	10	13.8	7900	91
	12	19.5	9400	159
	14	26.1	11000	237
Kavan 8 ×6	8	8.25	6670	43
	10	12.0	8180	76
	12	16.5	9680	142
	14	22.8	11090	205
	16	29.4	12600	311
Zinger (Wood) 7 × 6	10	7.8	8550	42
	12	9.6	10010	78
	14	12.3	11840	119
	16	15.3	13160	172
	18	19.5	15040	243
	20	23.1	16640	321

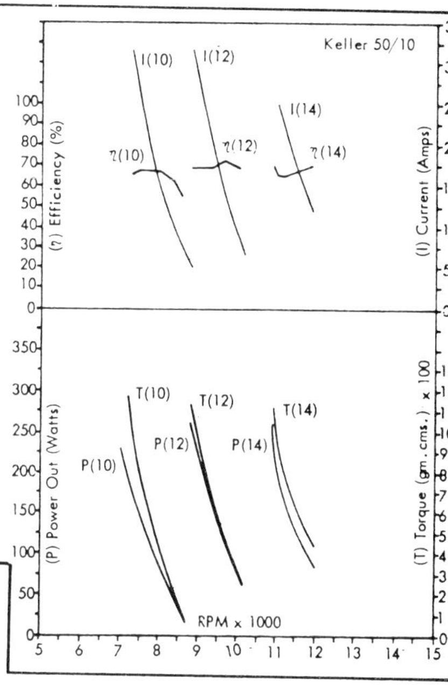

FIG. 19 Keller 50/10 motor performance.

KELLER 30/12

More powerful version of 30/8 for use with 12 cells. Weight 250gm. Recommended prop sizes 7in x 4in to 11in x 6in. (See fig. 18).

Photo 17. The Keller 30/8 and 30/12 motors.

KELLER 50/10

While intended for 10 cells, this samarium cobalt magnet motor can accommodate more and will turn props from 7in x 6in to 12in x 6in. Weight 400gm (See fig. 19).

KELLER 50/24

Substantially more powerful version of 50/10 suitable for up to 24 cells. (See fig. 20). Weight 395gm.

Photo 18. Keller 50, 50/10 and 50/24 motors.

Motor — Keller 50/24

Propeller	V_{in} (V)	I_{in} (A)	RPM	P_{out} (W)
Kavan 12 × 6	14	11.7	6200	115
	16	14.7	6960	172
	18	17.4	7710	232
	20	21.3	8220	285
	22	26.1	8930	365
Kavan 11 × 6	14	8.4	6390	108
	16	10.5	7050	120
	18	12.6	7900	158
	22	18.3	9310	280
	24	22.5	9960	345
	26	27.0	10430	418
Kavan 10 × 6	14	6.6	6670	62
	18	9.6	8460	117
	22	13.2	9870	190
	24	15.0	10530	227
	26	17.1	11280	287
	28	19.5	11840	328
	30	24.0	12220	358
Topflite (Wood) 9 × 6	14	6.3	6860	53
	18	9.0	8550	112
	22	12.0	10150	188
	24	13.8	10810	225
	26	15.9	11470	265
	28	21.9	11840	310
	30	24.6	12130	355
Kavan 8 × 6	14	5.4	6860	48
	18	7.8	8740	94
	22	10.5	10340	159
	24	12.3	11190	207
	26	13.2	11560	232
	28	15.6	12500	289
	30	19.2	12880	326
Zinger 7 × 6 (Wood)	18	5.4	9030	49
	22	6.6	10900	101
	24	7.5	11840	119
	26	7.8	12690	147
	28	10.5	13910	182
	30	12.0	14950	230

FIG. 20 Keller 50/24 motor performance.

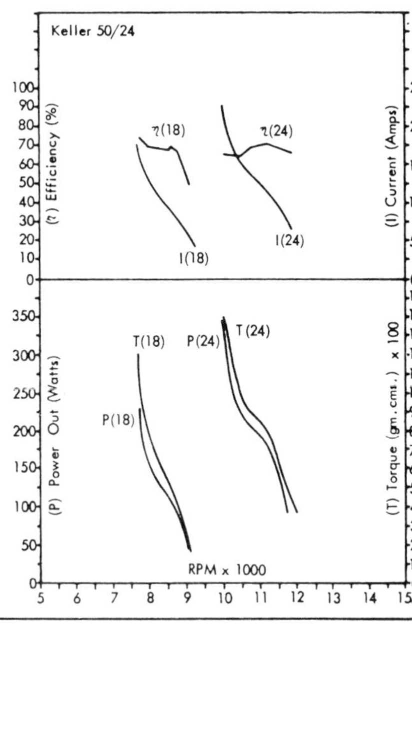

ROBBE ELT-MAX 30

Although similar in construction to the *Keller* motors, the Robbe motors do not employ samarium cobalt magnets and are, therefore, considerably cheaper. The Elt-Max 30 can operate from up to 20 cells with 8in x 6in to 11in x 6in props. Weight 380gm. (See fig. 22).

ROBBE ELT-MAX 50

Larger magnets and armature give considerably more torque than the Elt-Max 30 and at reduced current. Weight 560gm. (See fig. 23).

Motor — Robbe Elt-Max 30

Propeller	V_{in} (V)	I_{in} (A)	RPM	P_{out} (W)
9 × 6 Topflite (Wood)	12	11.7	6860	58
	14	14.1	7710	77
	16	15.6	7900	91
	18	17.4	8370	103
	20	19.5	8740	115
8 × 6 Kavan	12	10.5	7520	58
	14	12.9	8460	85
	16	15.6	9210	114
	18	18.3	9780	136
	20	19.8	10130	156
7 × 6 Zinger (Wood)	12	7.2	8740	47
	14	9.6	9780	75
	16	11.4	11090	103
	18	138	12030	130
	20	15.6	12970	160
7 × 6 Topflite	12	6.0	9590	44
	14	7.5	11090	60
	16	9.0	12220	75
	18	10.8	13350	103
	20	12.0	14380	133
	25	16.5	16730	206

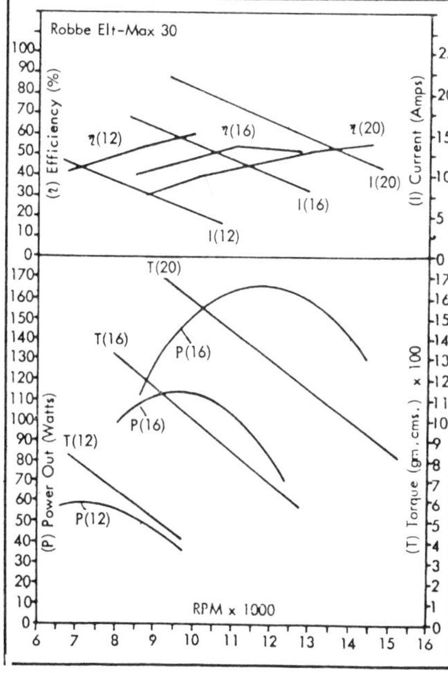

FIG. 22 Robbe Elt-Max 30 motor performance.

Mechanical details

Motor	Weight (gms)	Length x dia. (mm)	Shaft dia (mm)
Keller 30/8	250	62 x 40	5
Keller 30/12	250	62 x 40	5
Keller 50/10	400	74 x 42	5
Keller 50/24	395	74 x 42	5
Robbe Elt-Max 30	380	80 x 42	5
Robbe Elt-Max 50	560	105 x 42	5

FIG. 21 Keller and Robbe motor dimensions.

Photo 19. The Robbe Elt-Max 30 and 50 motors.

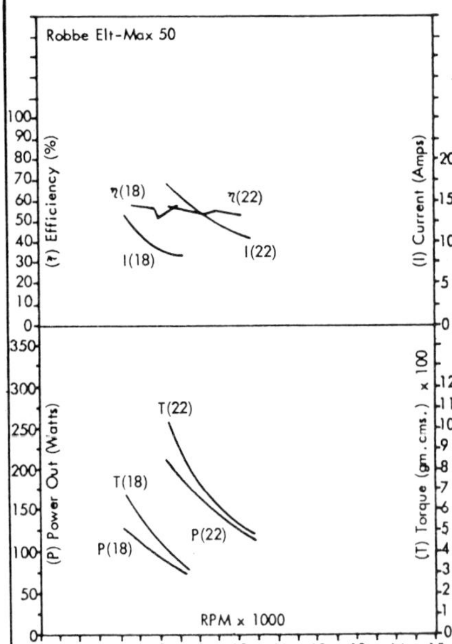

FIG. 23 Robbe Elt-Max 50 motor performance.

Motor — Robbe Elt-Max 50

Propeller	V_{in} (V)	I_{in} (A)	RPM	P_{out} (W)
Kavan 11 × 6	18	12.3	7240	128
	20	14.4	7900	164
	22	16.2	8370	200
	24	18.0	8550	217
	26	19.5	8930	241
Kavan 10 × 6	18	9.0	7900	91
	20	10.8	8550	119
	22	12.6	9210	149
	24	14.4	9870	183
	26	15.9	10240	205
	28	18.9	10900	260
Topflite (Wood) 9 ×6	18	9.3	7990	86
	20	11.1	8840	123
	22	12.9	9490	152
	24	14.7	10060	186
	26	15.9	11090	222
	28	17.7	11190	241
Kavan 8 × 6	18	8.1	8550	86
	20	9.3	9310	108
	22	11.1	10060	132
	24	12.9	10620	172
	26	14.4	11280	209
	28	16.2	11840	246

Chapter Four

Batteries

Batteries are the main source of direct current (D.C.) electrical power and fall into two main types:
a) Primary cells. These *manufacture* electricity by chemical means and, once exhausted, must be discarded.
b) Secondary cells. These *store* electricity and may be recharged and used many times. After several hundred cycles or misuse, there will be a gradual reduction of performance which, in practical terms, cannot be recovered.

While virtually any type of battery *can* be used to supply power to a motor, ranging from the simple zinc/carbon primary cell to the expensive and short-lived silver/zinc secondary cell, or even more exotic types, the most practical battery for electric flight purposes is the nickel/cadmium secondary cell.

NICKEL/CADMIUM CELLS

Known by many names (Ni-Cad, Ni/Cd, Nicad, DEAC, etc.) the nickel/cadmium battery is, perhaps, one of the least understood devices produced by mankind. One seldom appreciated reason for this is that there are many different ways of producing the cells, and each method produces different characteristics. Early cells were not capable of delivering high currents and, if overloaded, showed a dramatic voltage drop which effectively limited the available power. Any noticeable increase in cell temperature during discharge was an indication that damage had occurred and total failure could be expected within the next two or three charge/discharge cycles.

Modern high discharge cells can deliver very high currents without a large voltage drop and can become hot enough to cause burning of the skin without damage to the cell. However, repeated treatment of this kind will reduce the cell life.

A popular misconception is that the voltage of a nickel/cadmium cell remains virtually constant until the charge is virtually exhausted. In fact, the voltage of a fully charged cell is entirely dependent on the rate of charge. At normal charging rate (14-16 hours) the voltage per cell will be around 1.5 volts.At higher rates (see chapter 6) this can go to 2.0 volts or higher. When the charging current is removed, the voltage will immediately begin to fall and will stabilise at some point between 1.4 and 1.25 volts depending on numerous factors. This voltage cannot be taken as an indication of the health of the cell, nor can it be taken as an indication of *State of Charge.* A fully discharged cell will, after being charged for a short period, indicate the same voltage.

When current is taken from the cell, the voltage will again start to fall. This fall will continue until the load is removed, or the cell is exhausted (see fig 24). Initially, the voltage will fall rapidly,

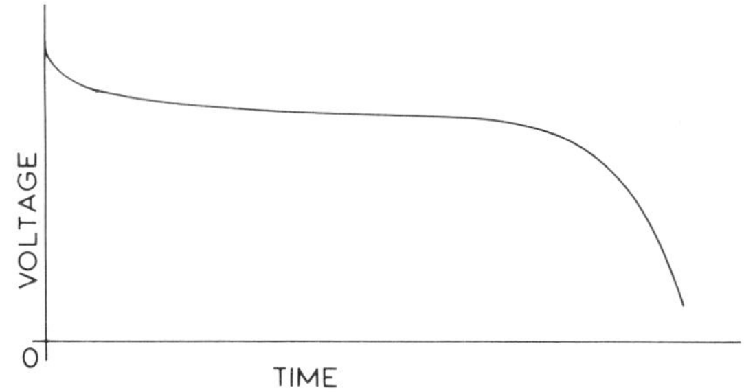

FIG. 24 Nickel cadmium cell discharge curve, constant load.

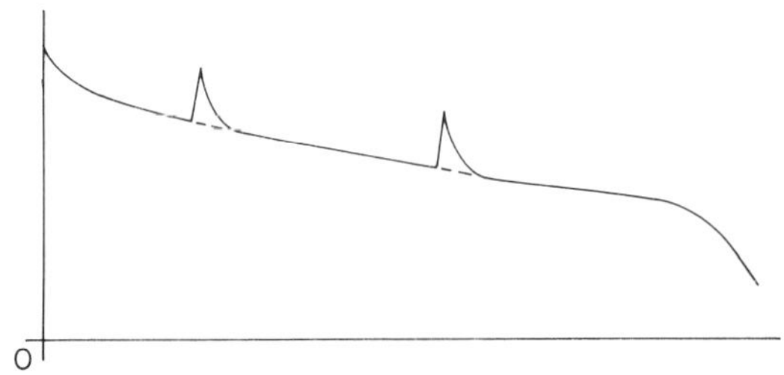

FIG. 25 Nickel cadmium cell discharge curve, interrupted load.

but the voltage/time curve will then become less steep until the charge is nearly exhausted, when there will be another sharp drop.

If the load is removed before the charge is exhausted, the voltage will immediately rise and again stabilise between 1.25 and 1.40 volts. Reapplying the load will produce a rapid loss of voltage until the original curve is resumed (see fig 25).

Thus, it can be seen that a healthy nickel/cadmium cell will have a no-load voltage between 1.25 and 1.40 volts *whether it is fully charged, partly charged or fully discharged* and the voltage is no indication of charge state.

STATE OF CHARGE

It may be assumed from the above that the charge state can be determined from the on-load voltage. This is true to some extent but requires that the tester is fully conversant with the particular cells being tested and with the testing equipment used and has a lot of experience in *interpreting* the results. Even then, the answer will only be an educated guess.

STATE OF HEALTH

A rough indication of cell *health* can be obtained by subjecting the cell to a high load and observing the voltage. The load should be at least equal to the cell capacity, i.e. 1.2 amps for 1.2 amp/hour cell. If the cell is in use regularly subjected to higher currents than this, then the test should be made at a higher current.

When the load is applied to a fully charged cell, the voltage should fall fairly slowly after an initial drop. Removing the load should cause a virtually instantaneous rise. A rapid drop and slow rise indicates that the cell should be replaced.

POLARITY REVERSAL

So far we have talked only of single cells. One modern cell can be discharged to zero volts without damage. If the load is then immediately removed, the voltage will rise to about 1.20 to 1.25 volts. Further discharge will produce an instantaneous drop to zero.

Normally cells are connected together in series (see fig 26). to form a *battery*. If a battery of cells is discharged to zero volts, it is highly likely that one cell will reach zero before the others. In this case, further discharge will cause that cell to reverse its polarity (in effect, it will acquire a negative voltage) due to being reverse charged by the other cells.

With some types of cell, this will do no damage, with others there may be loss of capacity or total failure. As a general rule, therefore, a battery of cells should *never* be discharged to zero volts. If this is felt to be necessary for some reason, the cells should be discharged *individually*.

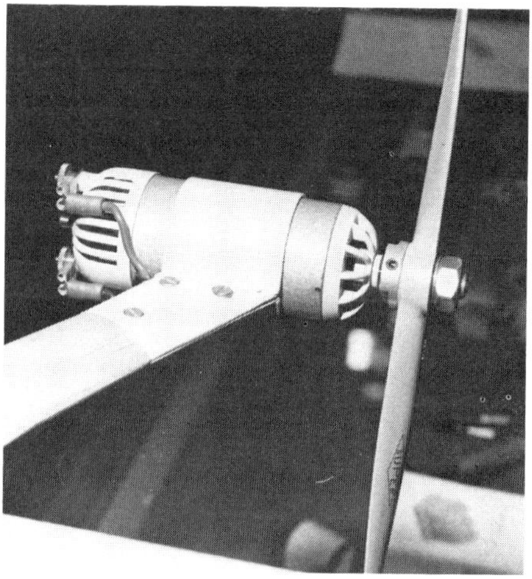

Photo 20. A Marx-Luder motor specially produced for electric flight.

DRIVING A MOTOR

When nickel/cadmium cells are used to drive a motor, the voltage will follow a curve similar to that already described. Since the current drawn by a given load is proportional to the available voltage and motor speed is roughly proportional to the current, the motor speed will follow a similar curve. At some point on that curve, there will be insufficient power to fly a given model. To obtain the maximum performance this point should coincide with the start of the rapid voltage drop at the end of the discharge curve. For general flying this means that the many variables should

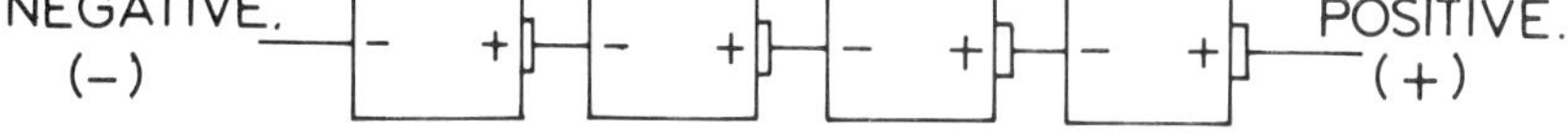

FIG. 26 Cells connected in series.

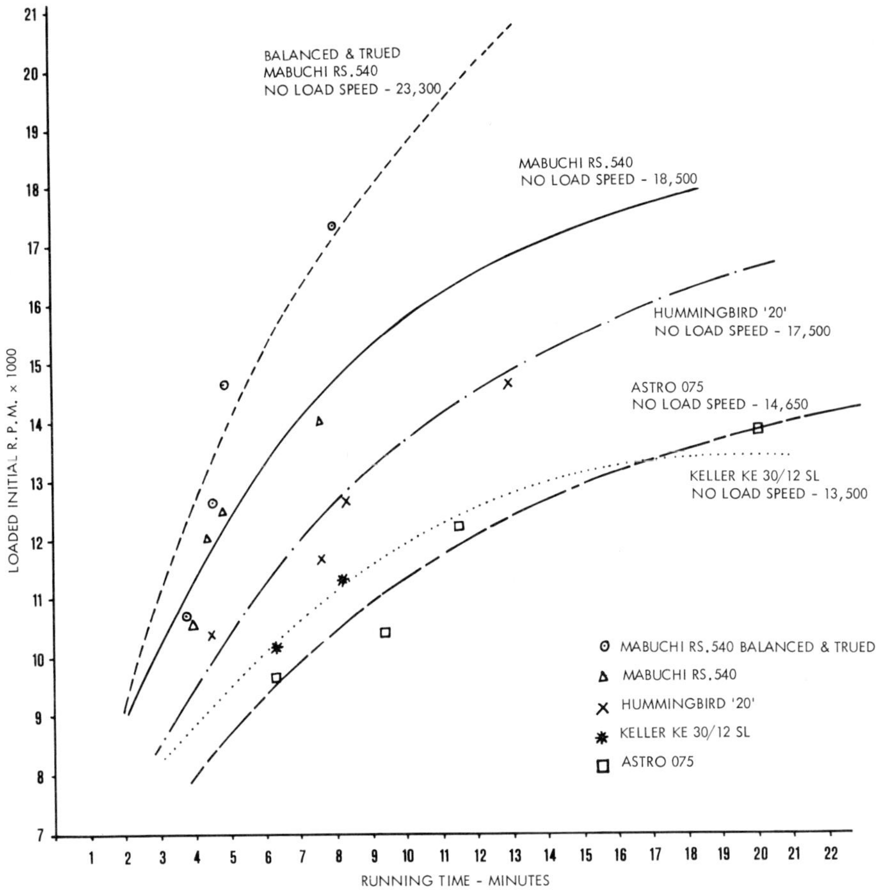

FIG. 27 Running time/RPM for various motors.

be arranged to give longest flight time.

One point worth remembering is that while motor speed is proportional to current this applies only to a *given load* (propeller size). Reducing the load will reduce the current for a given rpm or increase rpm for a given current. Therefore, the correct choice of propeller is the single most important factor of electric flight. Only trial and error can determine the optimum relationship of propeller/performance/flight time. (see fig 27).

Beware of using cells which are incapable of sustaining the current required by a given motor/propeller combination. Larger capacity cells may give much improved performance, despite increased weight.

SPECIAL USES

Where performance can be sacrificed for duration (i.e. powered gliders), the choice of cell type is non-critical, since loads will be modest and a typical flight will last until the charge is exhausted.

If duration is to be sacrificed to performance, then only high discharge, vented type cells should be used and care taken in charging, since cells may not be fully discharged (see Chapter 5).

CAPACITY

A cell's capacity is its ability to supply a certain current for a certain length of time. Thus a 1.2 Amp/Hour cell can, in theory, supply 1.2 amps for 1 hour. In practice, the manufacturer's rating is based on a 'ten hour rate' and it is more accurate to state that a 1.2 amp/hour cell can supply 0.12 amps for 10 hours.

When subjected to higher discharge rates, there will be a marked reduction in performance at some point before reaching the full capacity and the *apparent* capacity will be somewhat less.

At very high discharge rates, the apparent capacity may be only 60% of the rated capacity. Some types of modern high discharge cells can show considerable improvement on this figure.

Ageing cells will show a reduction of capacity. This is normally gradual and irreversible. Vented cells will, if overcharged, lose electrolyte through the vent, resulting in a sudden loss of capacity.

Capacity can only be checked at the 10 hour rate (i.e. 120 milliamps load on a 1200 milliamp/hour cell) and the result will have little relevance to electric flight use.

MEMORY

In some industrial or space flight applications where nickel/cadmium batteries are subjected to many hundreds of *identical* charge/discharge cycles, a phenomenon known as 'memory' can appear.

Where cells are repeatedly fully charged and then only 50% discharged, they will eventually only be capable of supplying 50% of a charge, despite being fully charged.

It is *extremely* unlikely that this effect could appear in any modelling application, and it may be safely ignored.

CYCLING

Where a battery *has* acquired a memory this can be removed by several full charge/discharge cycles.

This has led to the belief in some modelling circles that 'cycling' can restore the capacity of old, tired cells. In fact many 'name' manufacturers have never heard of cycling, while some insist that repeated 'deep' charge/discharge cycles can cause damage!

Tests show that an ageing battery of reduced capacity will show an increase after several cycles. However, when returned to normal duty, this 'extra' capacity will disappear during the next two charges.

EXTREME USAGE

When subjected to very high rates of charge/discharge, some types of cell produce marked characteristics. In some cases, a higher discharge rate can be achieved by using a very high charge rate. Some cells are damaged if allowed to become warm during charging, while

Photo 21. A Mercury Gnome *glider converted to electric power by the addition of a Mabuchi A1 unit over the C.G.*

others produce maximum performance if charged until they are quite hot.

Considerable research in this area has been done in the electric car racing world where it was recently rumoured that one team was using a set of cells just *once* after extreme charging methods and then throwing them away. The truth was that they were used once and then subjected to a thorough testing and recharging by the team's battery 'technician' before being subjected to further brutal 'topping up' and use.

This in fact was a wise precaution, since drastic overcharging of *any* type of nickel/cadmium battery can be *very dangerous.*

WATER ACTIVATED CELLS.

In the early days of electric powered F/F model aircraft, a form of cell manufactured for life-saving equipment was used. The chloride depolarised water-activated cell came in many forms from several manufacturers, some needing plain water, others salt water, to activate them.

They could deliver surprisingly high currents for short periods (typically 0.3 amps at 2.8 volts for 45 minutes from 1oz weight), but demand far outstripped supply, and they became rather difficult to obtain. Being primary cells they could be used only once.

LITHIUM BATTERIES

Another type of primary cell which has enormous possibilites is the lithium battery. For a given weight, they have roughly 6 times the capacity of a nickel/cadmium cell. Unfortunately, they are very expensive, can only be used once, and are at the moment rather limited in the current they can supply. This situation should improve with more development.

CYCLON CELLS

A battery which could be of use to us is the *Chloride Gates* 'Cyclon' cell, which is in fact a sealed lead/acid battery. It has slightly less energy for a given weight than the nickel/cadmium cell. Voltage per cell is 2.0 volts.

SILVER/ZINC BATTERIES

These have considerably more power for a given weight than the nickel/cadmium battery, but they cannot give a high output current, nor can they be fast charged. They are very expensive.

SPECIFIC OUTPUTS

For a battery weighing one pound, the output in watt/hour for each type cell will be as follows:

Chloride Gates 'Cyclon sealed lead/acid	11.5
Nickel/Cadmium	13
Silver/Zinc	50
Lithium	120

PREVENTING REVERSE CHARGING

One method of preventing reverse charging is to connect a diode across each cell so that a reverse voltage will be shorted out. (see fig 28). This practice is widely used in receiver and transmitter packs on quality R/C equipment.

To be fully effective, the diode should be able to withstand the highest current likely to be passed into or out of the battery. In electric flight operations this could require the use of very large diodes.

Even so, it is likely that the diode will not be required until the battery is almost fully discharged, when the current may have fallen to a low level. Since diodes in 1 amp and 3 amp ratings are readily available and very cheap, the fitting of them is well worth considering.

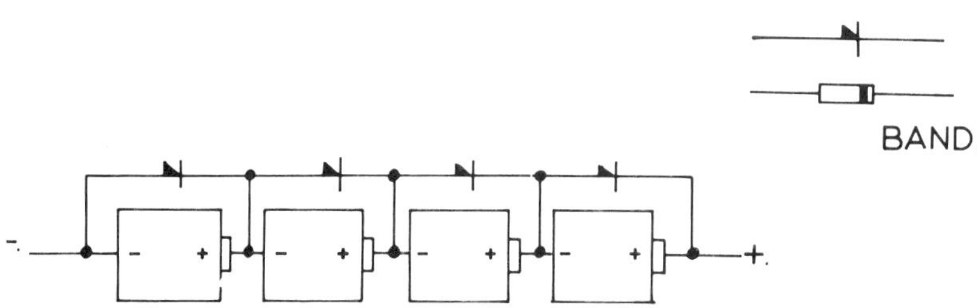

FIG. 28 Fitting diodes to prevent polarity reversal.

Chapter Five

Charging

Having decided that nickel/cadmium batteries are currently the most practical source of power, it becomes necessary to find some means of charging them.

There are two main power sources available for charging:

a) Mains electricity

b) Large capacity batteries.

CHARGING FROM MAINS ELECTRICITY

The one advantage of mains electricity charging is that you do not need to buy and maintain the large capacity batteries which would otherwise be required as a charging source.

However, apart from the obvious disadvantage of lack of portability, mains charging at high current is very expensive due to the need for a large transformer.

In fact it should be possible to buy a charger designed for 12 volt operation, a 12 volt car battery *and* a car battery trickle charger for about the same price as a mains powered charger of similar performance.

Do not be misled into thinking that a car battery charger alone is a suitable method of charging. Their output is far from smooth and D.C. and Automatic chargers will not function satisfactorily, while simple resistance charging can give misleading results.

CHARGING FROM BATTERIES

Here the possibilities are endless. Small lightweight systems (like the Mabuchi A.1 unit) can be charged from zinc/carbon batteries, while various lead/acid or high capacity nickel/cadmium cells of ex-government or industrial use can be used for charging more sophisticated systems.

Undoubtedly, the most common battery used is the 12 volt lead/acid battery, be it a car type, motor-cycle type, or one of the sealed 'gel-cell' type.

One problem here is that a maximum of eight nickel/cadmium cells can be charged from 12 volts, while seven is more practical. Therefore, it may be necessary to use two or more 12 volt batteries, or divide up your aircraft battery and use staggered charging methods (see fig. 29).

One point which should be mentioned here is that nickel/cadmium batteries should *never be connected in parallel,* so don't attempt to charge two 6 cell packs simultaneously from one 12 volt battery, unless taking suitable precautions (See fig. 30).

The actual means of charging to be employed will depend on the number of cells to be charged and the charging time required. For electric flight purposes it is not necessary to use high rates of charge and a convenient charg-

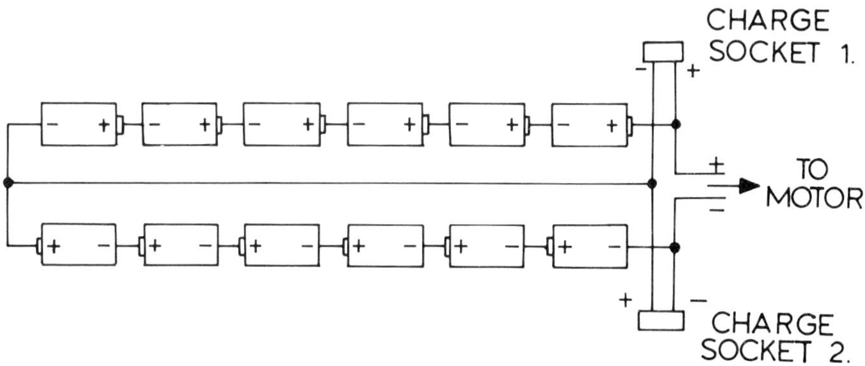

FIG. 29 Method of charging 12 cell pack as two packs of 6 cells.

ing current is twice the rated cell capacity, i.e. 2.4 amps into a 1.2AH cell.

TIMED CHARGING

It might be assumed that this would give a charge time of 30 mins:

$$\frac{1.2AH}{2.4\ A} = .5H$$

However, we have already seen that at high discharge rates a cell will not give its full rated capacity. The same is true when charging. Fortunately, considerable experience has been gained with 1.2AH cells and this indicates that a safe figure for charging at 2.4 amps is 64 amp/minutes.

This gives a charge time of:

$$\frac{64}{2.4} = 26.66 \text{ minutes}$$

Which makes it possible to use a timed charge and gain all of the usable capacity.

The problem here of course is that you *must* ensure that the cells are fully *discharged* before charging.

Experience indicates another problem

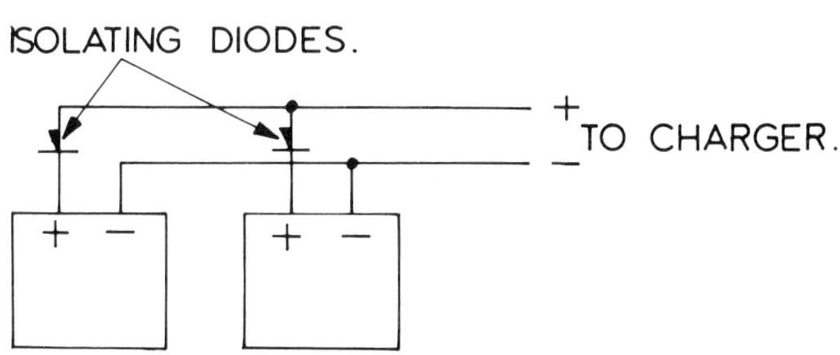

FIG. 30 Only possible method of charging two packs in parallel.

– either use some type of mechanical or electronic method of terminating the charge or at least use some type of alarm – alarm watches are now obtainable, cheaply. Merely looking at a clock when you start charging *will not do.* You will eventually forget to watch the clock and will ruin a set of batteries!

Timed charging is perfectly satisfactory for up to 6 cells from a 12 volt source. If seven cells are used, you must ensure that all connections are kept in good order and use short, thick, wires to minimise voltage drop. The current will drop towards the end of the charge as the surface voltage of the cells rise, and it will be necessary to use some interpolation to guess-timate the average current and adjust the charge time accordingly. Always err on the side of caution.

When charging eight cells, the charge time can only be determined by experience and any variation in connections, 12 volt battery condition, state of charge, etc. will influence this.

If charging from the battery installed in your car, remember that charge rates and times will be considerably different if the car engine is running. Likewise different results will be obtained between connecting charge leads directly onto the battery terminals and plugging into the car cigar lighter socket. This is due to the additional wiring involved.

In short, always charge in the same way, under the same conditions, and change the charge time to suit, as dictated by experience.

Photo 22. Charging on the field. This Electra-Fli *is using four 6v scooter batteries through a time-set charger.*

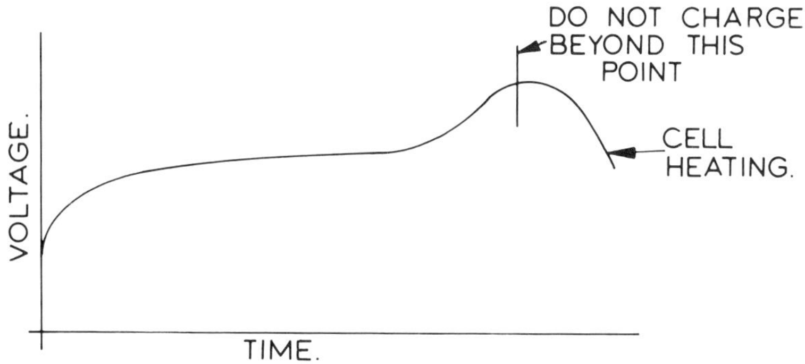

FIG. 31 Typical charging voltage curve.

CHARGING BY VOLTAGE MONITORING

A useful characteristic of nickel/cadmium cells is that, when almost fully charged, the rate of change of voltage increases. (see fig. 31).

This change can only be detected with very sensitive equipment, a digital voltmeter (DVM) being ideal.

It is essential that the meter is carefully watched during the charge, since it is necessary that the normal rate of change be determined in order to recognise an increase in the rate.

Under normal conditions it is best to terminate the charge when the rate of

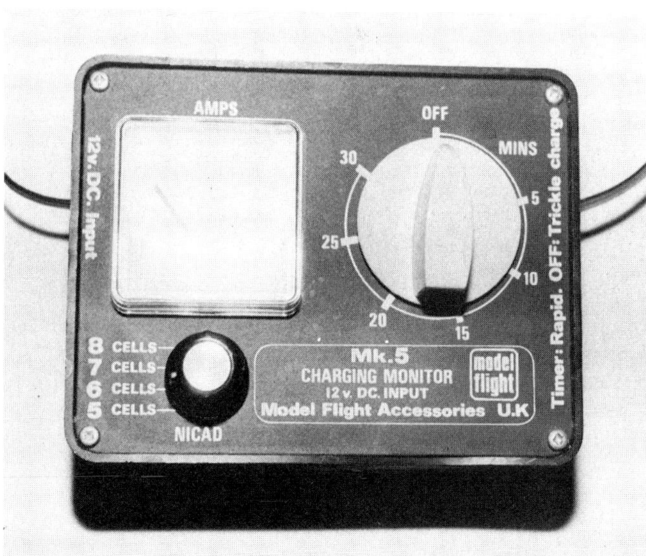

Photo 23. An MFA charger with time control, ammeter and switching system for 5,6,7 or 8 cells. Reverts automatically to trickle charge.

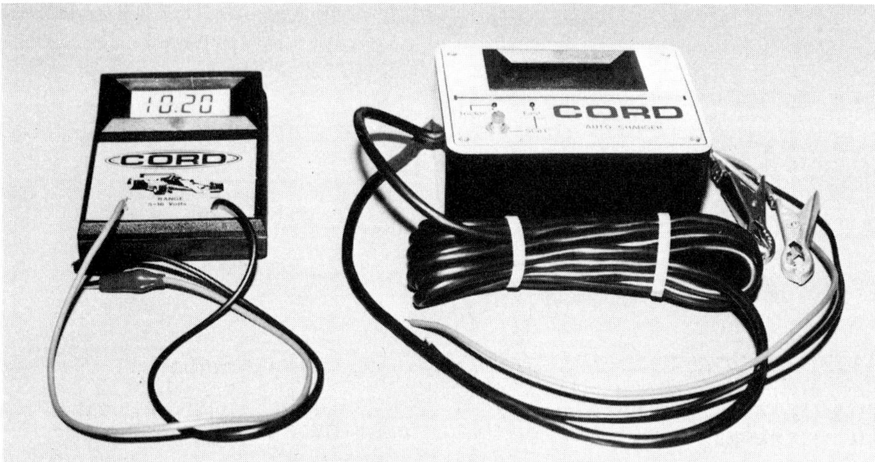

Photo 24. Cord equipment – a digital voltmeter and automatic charger which switches to trickle charge when full fast charge is achieved

change begins to increase. For special conditions i.e. contest flying, stop charging when the voltage stops rising and is about to start falling.

The advantage of this method is that it will work under any conditions for any size of cell regardless of charge rate, number of cells, state of charge, etc.

While a DVM is an expensive item, it is well worth acquiring one.

AUTOMATIC CHARGERS

There are several of these currently available, ranging from simple mechanical clock types (see photo. 23) up to electronic types capable of sensing the increase in rate of change voltage. Most of the latter are designed for the model car world and cater specifically for charging six cells.

Some of these automatic chargers, notably the *Cord* (see photo. 24),, give details for use with more or less than six cells.

Automatic chargers can also be obtained which are designed to stop charging at a certain voltage (this may be preset or adjustable). This can be satisfactory *providing* the cells are *always* charged in the same manner and no attempt is made to charge to the limit of their capacity. As already noted, the fully charged voltage of a cell or cells is a variable depending on numerous factors.

SLOW CHARGING

So far we have only considered rapid charging. It is advisable occasionally to give cells an overnight charge at the normal rate of one tenth of the rated capacity i.e. 120 mA into a 1200 mAH (1.2AH) cell.

It is often claimed that this gives increased capacity but it is doubtful whether this is the case (See fig 32).

TYPES OF CHARGING

The charge rate is normally given in

Mabuchi RS 540 motor with 7in. x 6in nylon prop (Tornado)
Running time in minutes. RPM taken at start of run

Eight cells 1.2 Ah	Battery 1		Battery 2		Battery 3	
	RPM	Running Time	RPM	Running Time	RPM	Running Time
Overnight charge (14 hours at 120mA)	10,300	3.50	10,400	3.50	10,700	4.50
DVM monitored charge (two amps)	10,300	4.05	10,200	4.08	10,700	4.30
20 minute charge (car battery)	10,500	3.55	10,400	3.55	10,400	4.25

terms of capacity (C) and for *overnight* charging a figure of C/10 is normally used, this being the rate at which cells can be charged *continuously* without damage.

Rates between C/10 and C are known as *rapid* charging, while rates in excess of C (i.e. 2C, 3C, etc.) are known as fast charging.

When either rapid or fast charging are carried out, it is *essential* that cells are not overcharged. This will lead to overheating, venting and loss of capacity. It is widely held that the older, sealed, button cell, type of nickel/cadmium battery cannot be either rapid or fast charged. This is not true, but, *any* overcharging will lead to physical damage of the cell accompanied by an explosion!

CHARGING DO'S AND DON'TS

Don't: Overcharge
Connect cells in parallel
Give a timed charge without an automatic cut-off or alarm
Charge from a car with the engine running
Charge from a different source and expect the usual results.

Do: Fully discharge batteries before charging
Give an overnight charge occasionally
Use a DVM when available
Charge in exactly the same manner every time
Take, and refer to, lots of notes.

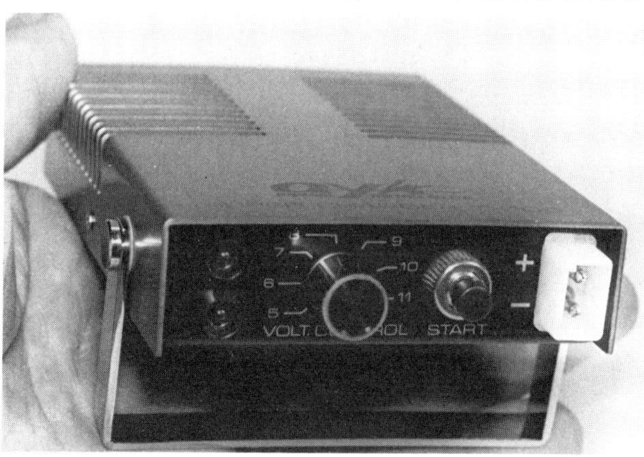

Photo 25. AYK charger with voltage adjustment and warning light system which changes when full charge is reached.

Chapter Six

Commercial Systems

This chapter will only concern itself with complete motor/battery systems.

ASTRO-FLIGHT.

The range of motors produced by Astro-Flight Inc. has already been covered in chapter 3. Each of these is normally supplied with a suitablc battery pack to produce a complete system.

02 F/F AND R/C SYSTEMS

Two different battery packs are available, a 250 mAH type for F/F and a 550 mAH type for R/C. For F/F purpose, the 250 mAH pack is normally given a timed charge to produce a limited run (typically 25 seconds), although the full run of around two minutes could be used for lightweight R/C purposes.

Flight times of around five minutes are available with the 550mAH unit. Recommended propeller in each case is a 5¼in x 3in *Topflite* nylon. (see fig. 33). Total weight 173gm or 290gm.

05 SYSTEM

Fig. 34 shows the flight performance of

Astro 020 RIC and Free Flight Systems

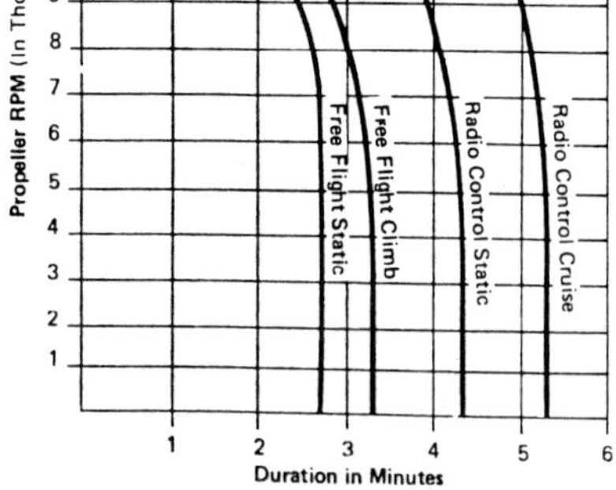

FIG. 33 Astro 02 system performance.

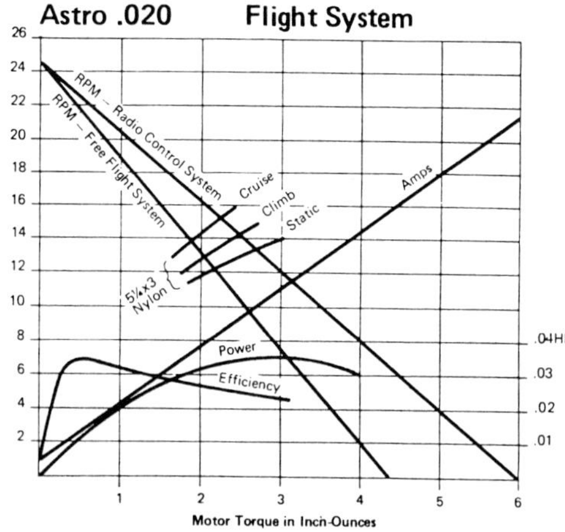

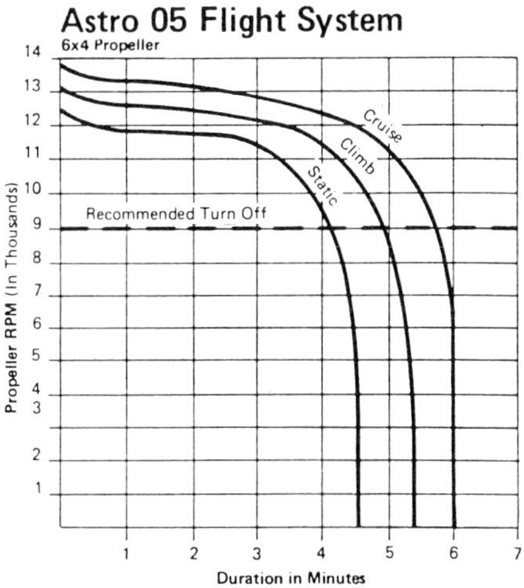

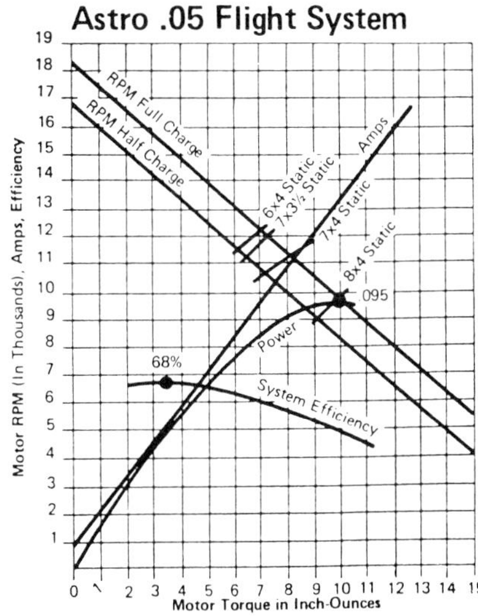

FIG. 34 Astro 05 system performance.

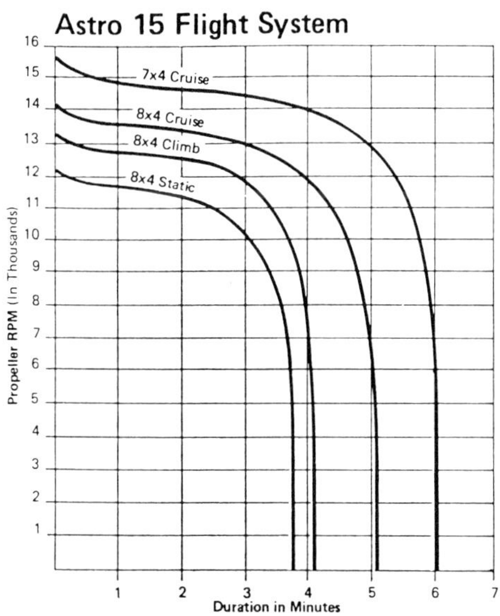

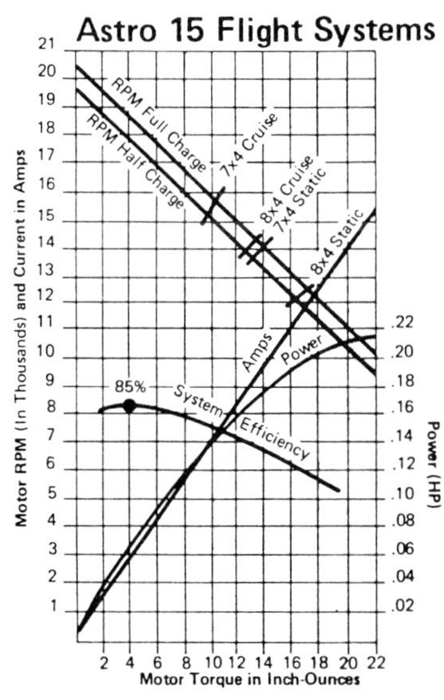

FIG. 35 Astro 15 system performance.

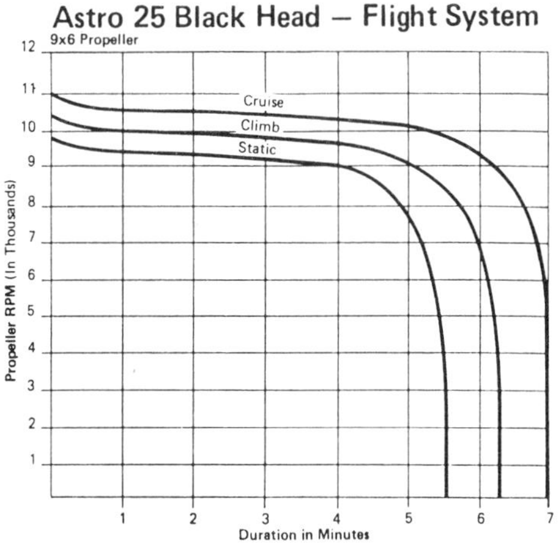

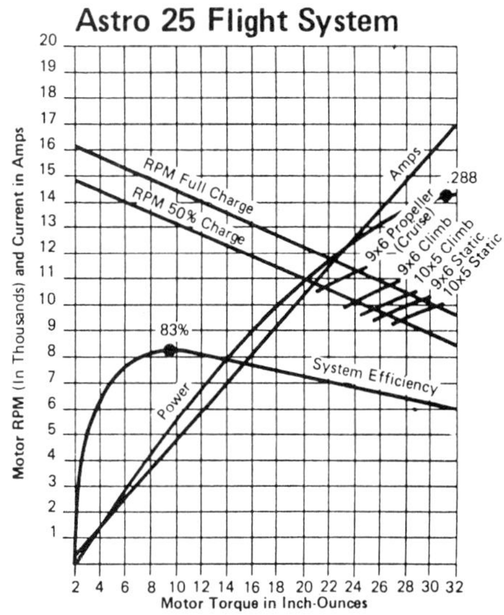

FIG. 36 Astro 25 system performance.

the 05 system when fitted with a *Cox* 6in x 4in propeller. Total weight 567gm.

15 SYSTEM

Recommended propeller is an 8in x 4in, but a 7in x 4in will give longer flights where smaller models are used (see fig. 35). Total weight 734gm.

25 SYSTEM

Flight performance when fitted with a 9in x 6in propeller is shown in fig. 36. Total weight 1183gm.

LEISURE FLIGHT

This company was started by one of the founders of *Astro-Flight Inc.* and its products are not well known in the U.K. Their 05 system uses a 'tuned' (i.e. rewound) motor giving better performance, but restricted to 6 cell battery packs and small propellers to avoid overheating problems. Some astonishing claims are made of this system. (See page 84).

MODEL FLIGHT ACCESSORIES (MFA).

This company has done much to promote Electric Flight in the U.K. and is probably responsible for the current state of the art.

'HUMMINGBIRD 02'.

Intended for small R/C or F/F models of about 16-20oz (450-560gm) all up weight. Recommended propeller 5in x 3in. Consists of RS380 type motor with five 600 mAH cells. Total weight 225gm.

'HUMMINGBIRD 05'.

An RS540 type motor is connected to eight 600 mAH vented cells. recom-

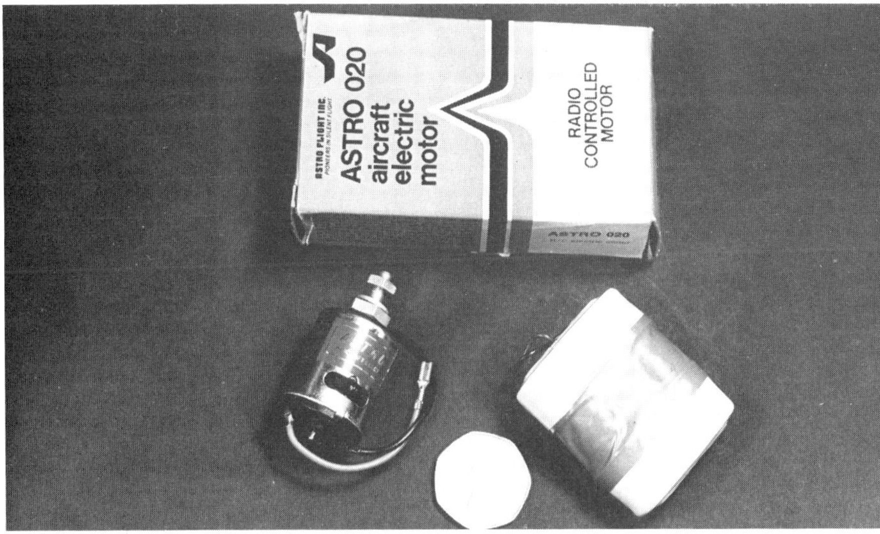

Photo 26. The Astro 020 with battery pack for R/C flying. Weight with the larger option battery is about 10 ozs for near 5 mins power.

Photo 27. Astro-Flight Inc. were one of the early firms in electric flight. Range of motors shown here is 25, 15, 10, 050 and 020.

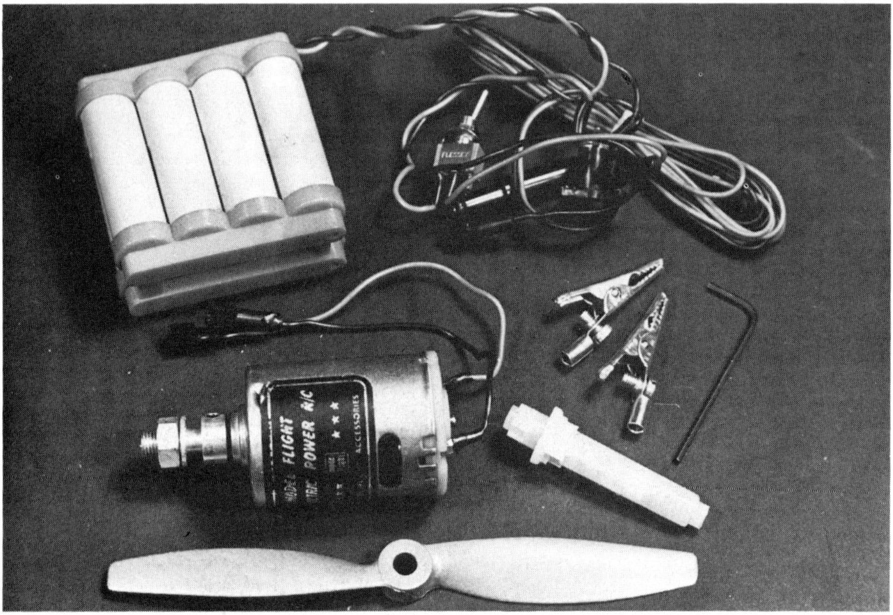
FLIGHT
POWER R/C
ACCESSORIES

Radio Models

Opposite, top. Photo 28. Among early British suppliers were MFA, whose Hummingbird 15 system is shown here. Note tubular fuse holder.
Opposite, bottom. photo 29. The attractive Hummingbird aircraft designed round the Hummingbird 15 system being charged from a 12v accumulator.
Right, photo 30. The Astro 05 motor and flight batteries, the most popular size for powered gliders and sport models.

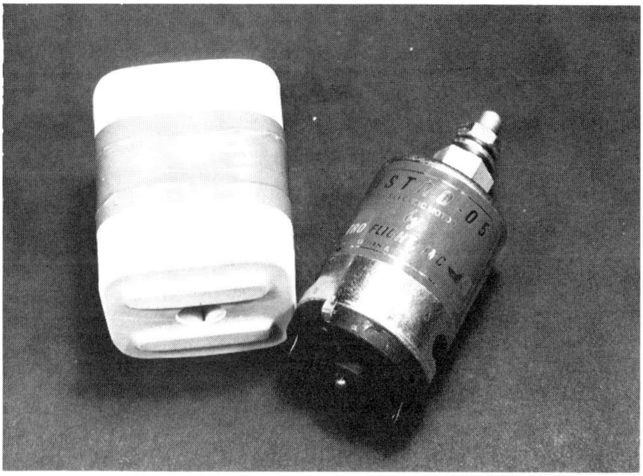

mended propeller (supplied) is a *Cox* 6in x 3in. Total system weight 385gm. Suitable for the MFA 'Hummingbird' aircraft (44in span, all up weight 0.95-1kg) or similar models.

TWIN POWER PACK

Here *two* RS540 type motors are powered by eight 1.2AH cells. Otherwise generally as the 05 system. Total weight 767gm.

'HUMMINGBIRD 15'.

Using the same battery as the twin power pack driving a single RS540 type motor, this unit will turn a 7in x 4in propeller at 11000 RPM with a duration of about five minutes. Total weight 600gm.

'HUMMINGBIRD 20'

Generally similar to the 15 unit but with the higher torque RS550 type motor. Recommended propellers, 7in x 6in, 8in x 4in or 8in x 6in. Total weight 600gm.

'HUMMINGBIRD 30'

With an uprated motor running on 16 cells of 1.2AH capacity, this unit will turn propellers of 8in x 6in, 9in x 4in, 9in x 6in or 10in x 4in size. Duration about five minutes. Total weight 1175gm.

WESTON UK

With a Mabuchi RS540 motor ready connected to eight selected Saft 1.2AH nickel/cadmium cells and complete with switch, charge socket, charge leads, propeller adapter and 7in x 4in propeller, the *Weston (UK) Ltd* unit is a very convenient way to produce your first electric model. All wiring is kept short to avoid power loss and RF interference. Total weight 600gm. (see photo. 31).

RIPMAX CYCLONE 15 SYSTEM

This unit has had Cyclone 15 motor coupled to an eight cell 1.2AH battery pack and includes wiring, switch, fuse and holder, shaft adapter and charging

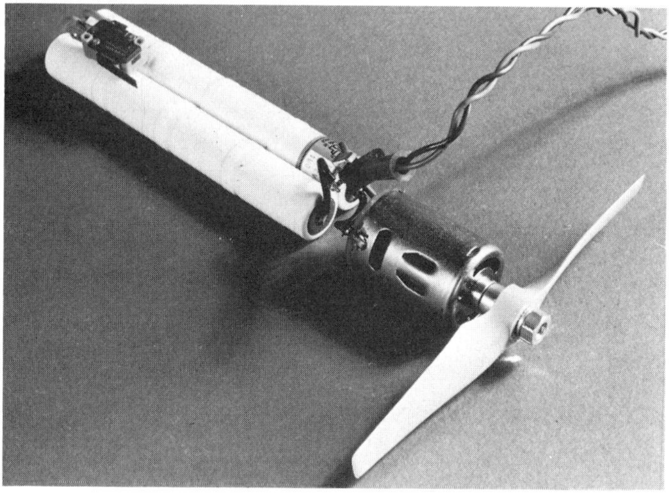

Photo 31. The Weston UK flight pack comes ready-wired and uses a standard RS.540. Turns a 7 x 6 in. prop at 10,000-11,500 rpm.

lead. (see photo. 32). See chapter 3 for motor details. Total weight 600gm.

RIPMAX BULLET 30 SYSTEM

Here the Bullet 30 motor is supplied with *two* eight cell 1.2AH packs and similar accessories to the Cyclone 15 systems, which produces a very powerful unit. (see photo. 33). Total weight 1260gm.

MABUCHI A.1 UNIT

Intended for F/F models, this is a com-

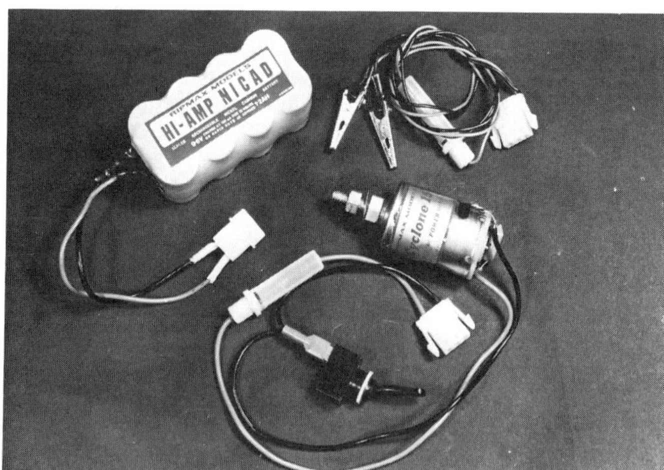

Photo 32. Ripmax Model's Cyclone 15 system was a popular outfit but it is believed that Ripmax may discontinue supply.

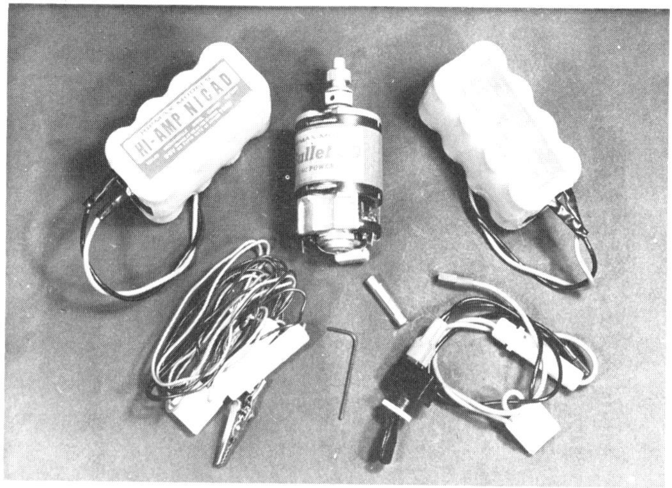

Photo 33. The Ripmax Bullet 30 system may also be discontinued but no doubt second-hand examples will be around for a long time.

plete self-contained unit featuring a small *Mabuchi* motor fitted with a 4½in propeller with a plug-on battery holder containing two 100mAH fast charge nickel/cadmium cells. A novel switch is included which is operated by rotating the motor relative to the battery holder.

For charging, the two cells are removed from the holder and inserted in a charge unit containing four U2 type zinc/carbon cells. A pushbutton on the charge unit connects the two cells to the zinc/carbon battery. Holding the button down for five minutes gives a full charge

Photo 34. Another free-flight pylon application of the Mabuchi A1 unit. Excellent for small-field flying and experimenting with shapes and weights.

Photo 35. The Acoms flight unit- note pinion on motor shaft. Gears allow motor to run at peak efficiency and permit reduction in battery weight.

and produces a motor run of 30-45 seconds. (See photo. 8). Total weight 64gm.

ACOMS UNIT

Designed for use with the *Acoms* 'Cessna' ARTF model, where it drives a 9in diameter propeller through reduction gearing, the *Acoms* motor and battery pack are available as spares and can form the basis of a system similar to the *Astro-Flight* 02 or *MFA* 'Hummingbird' 02 systems.

The motor is a *Mabuchi* RS380 type and the battery pack has five 600 mAH *Saft* cells. (See photo. 35). Total weight 290gm.

Chapter Seven

Connecting the Electrics

BASIC CONNECTIONS

Reduced to its simplest possible terms, we need only an electric motor and a battery (see fig. 37).

However, practical considerations require that we add a switch and some means of charging the battery (see fig. 38).

A stalled motor with a fully charged battery could lead to a damaged motor or even a fire, so we also need a fuse to disconnect the motor in the event of an overload (see fig. 39). A normal 13 amp mains fuse can be used for most systems.

If it is required to switch the motor on and off in flight, the switch could be op-

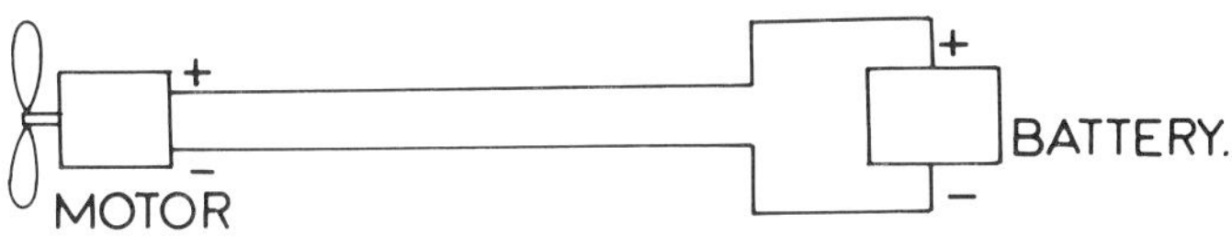

FIG. 37 Basic motor/battery circuit.

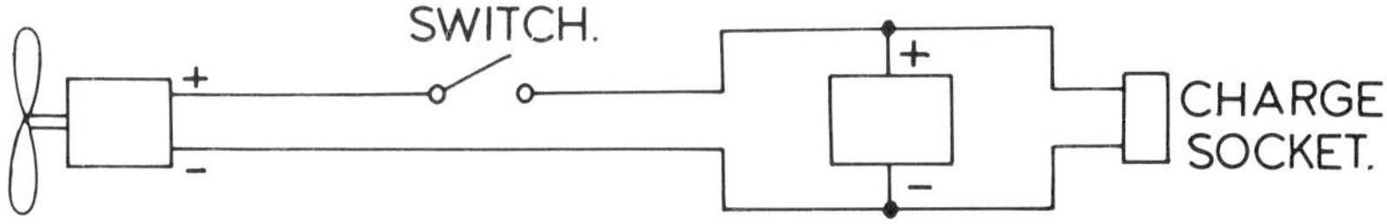

FIG. 38 Minimum practical circuit.

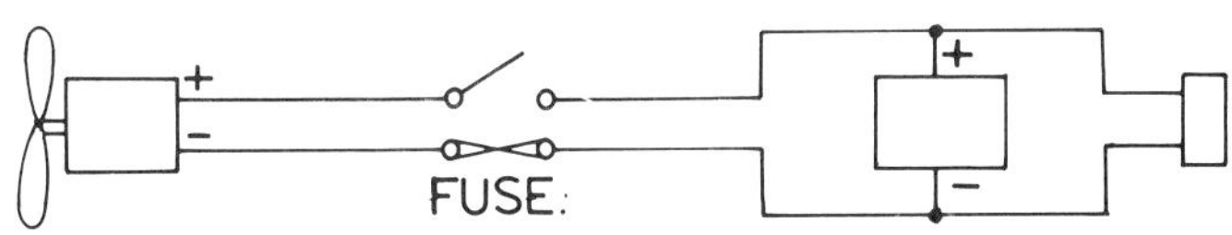

FIG. 39 Safe minimum circuit.

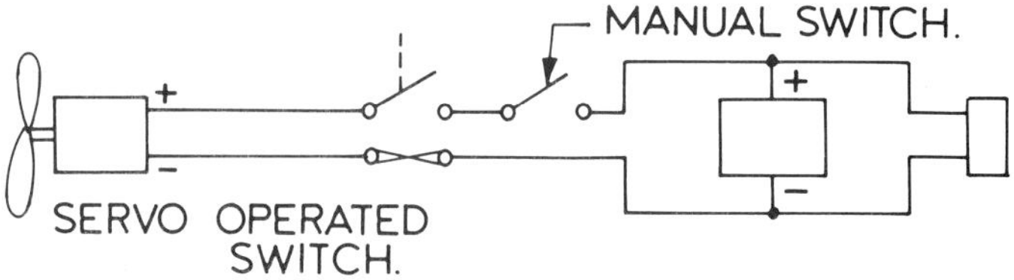

FIG. 40 Servo-operated switch with manual over-ride.

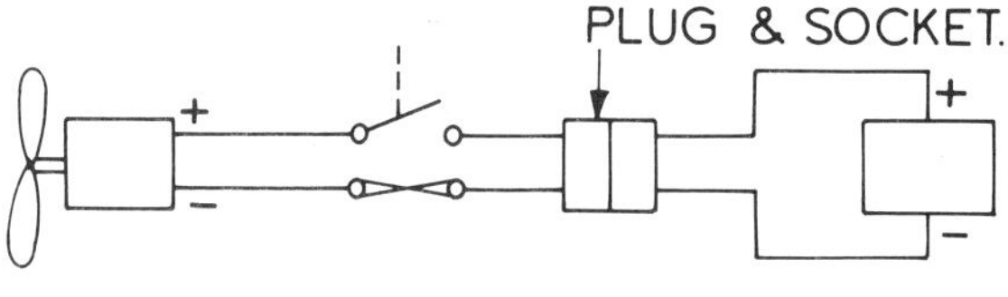

FIG. 41 Simpler version.

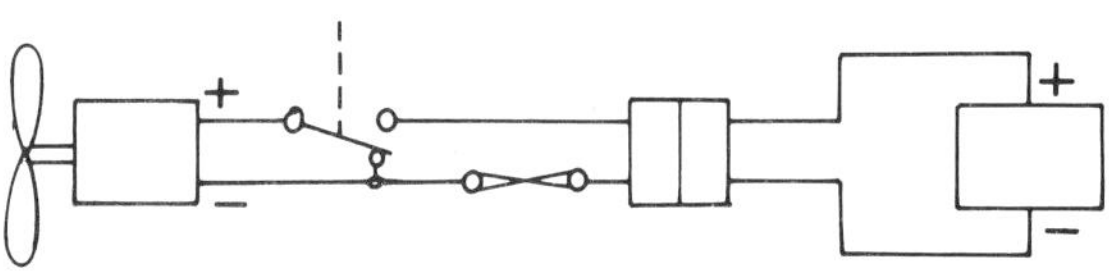

FIG. 42 Adding motor braking.

erated by a servo. In this case, it might be felt necessary to add another switch to prevent accidental switching on of the motor. This switch should be connected in series with the servo operated switch (fig. 40).

An alternative system is to have an easily accessible plug and socket between the battery and motor which can serve a dual function of manual on/off switch and charge socket (fig. 41).

One further addition to this is to use a change-over switch and use the other contact to apply braking to the motor to prevent the propeller from free-wheeling when the motor is switched off (fig. 42). This provision is not normally required with motors fitted to reduction gear units.

MULTI-CELL SYSTEMS.

Where more than eight cells are used, it will normally be necessary to charge the battery in sections. One method of doing this is shown in fig. 43. A better method appears in fig. 44.

LIGHTWEIGHT SYSTEMS.

One very obvious way of reducing the weight of an electric R/C model is to use the drive battery to power the R/C equipment. While this is established practice in the model car world, there are obvious hazards in aircraft use. The drive battery must never be discharged to the point where it interferes with the R/C equipment.

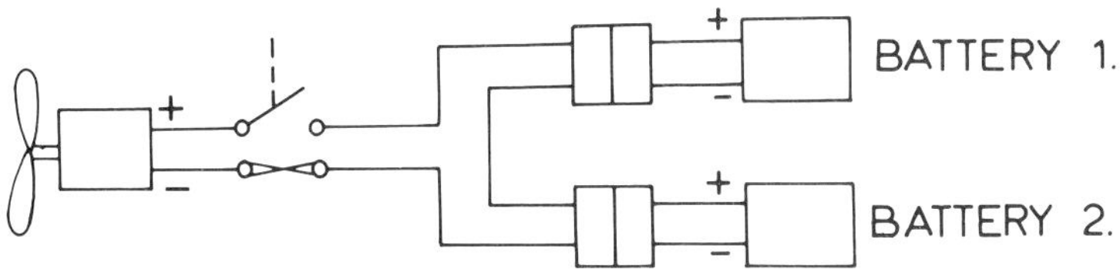

FIG. 43 Two half battery packs to facilitate charging.

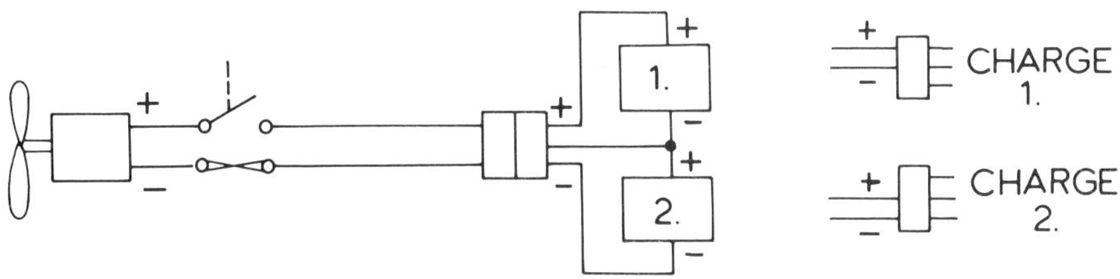

FIG. 44 Alternative method.

With multi-cell systems this is not such a problem. There are available many 3 pin, 5 volt integrated circuit voltage regulators which can be connected as shown in fig. 45, so as to provide a stable 5 volt supply to the R/C gear. Such devices commonly need differential of 1-2 volts for satisfactory operation. This means that there will be no interference to the radio until the drive battery drops to 6-7 volts.

When using drive batteries of 12 or more cells (nominally 15 volts) it is likely that the model will have stopped flying long before this point is reached and control is affected.

However, with 7 and 8 cell packs, there are two alternatives:-

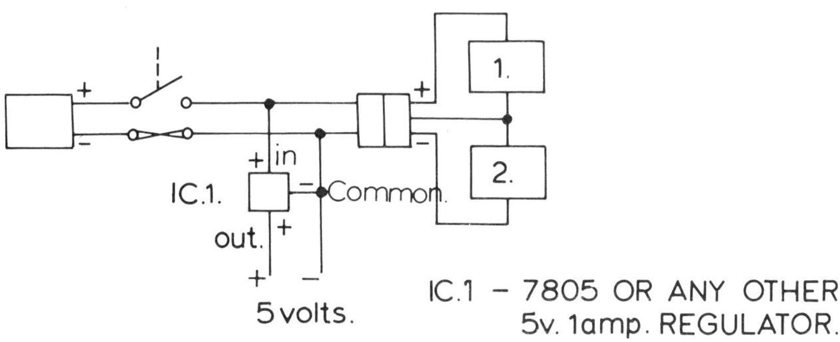

FIG. 45 Deriving a receiver supply from the drive battery.

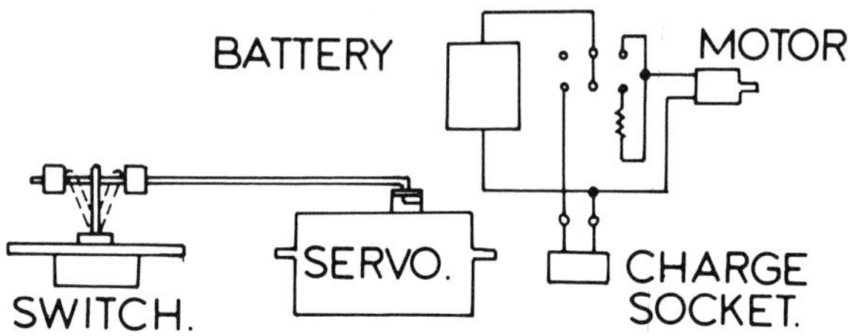

FIG. 46 Astro-Flight two-speed switch.

a) Make up a receiver battery pack from 100 mAH fast charge cells such as those supplied for use with the *Mabuchi* A.1. unit and charge them at the same time as the drive battery.
b) Arrange some method of stopping the motor at a pre-determined voltage.

AUTOMATIC SHUT-OFF

Fortunately, both *Astro-Flight* and *Leisure Flight* produce electronic packages designed to supply a stable voltage for the receiver and shut off the power before problems occur. They also incorporate a means of turning the motor on and off from an R/C channel.

For home constructors, a similar device, designed by the author, appeared in the March 1982 edition of *Radio Control Models and Electronics* magazine. (See appendix).

SPEED CONTROLLERS.

Speed controllers fall into two basic types:-
a) Resistance types. These reduce power by dissipating excess voltage in the form of heat. This is wasteful and not really suitable for aircraft.
b) Electronic types. Much development has gone into these for use in electric cars. Unless fitted with a full power over-ride relay they give a voltage drop which, when driving a propeller, can amount to a loss of 1500 RPM. When a full power relay is fitted, they are fairly heavy. They are also quite expensive.

For aircraft use, a speed controller is an unnecessary luxury, a simple on/off switch being quite sufficient for most purposes.

One exception for this is for scale models, which may be required to taxi. Here a simple two-speed resistance device may be satisfactory.

A suitable device is available from *Astro-Flight* (see fig. 46).

SERVO OPERATED SWITCHES

Usually the movement required to operate a switch is considerably less than that available from a servo. Some ways of overcoming this are shown in fig. 47.

This arrangement can be used with 2-channel radio equipment by operating this switch from the elevator servo with full 'up' turning the motor on and full 'down' turning the motor off. With the *Acoms* 'Cessna' it is possible to relocate the switch inside the fuselage and use this method.

SUPPRESSION

Electric motors all produce RF noise to varying degrees. Most commercial systems will have some means of suppression already fitted to the motor, this usually taking the form of a 0.01 to 0.047 microfarad capacitor connected across the motor terminals.

This will usually be sufficient, but some cases may require further measures. Connecting additional capacitors from each terminal to the body of the motor or earthing the motor body to the negative lead can be successful in specific cases.

Where the motor is switched by a relay, the contacts should have a 100 ohm resistor and a 0.047 microfarad capacitor mounted in series across the (see fig. 48).

All connections between motor and battery should be as short and as direct as possible and kept well away from radio equipment and, particularly, the receiver aerial. Some reduction of range with the motor running is normal.

REDUCING LOSSES.

In order to ensure maximum efficiency and, therefore, performance, all direct connections between battery and motor should be in good condition and adequately rated. Wiring should be kept short and as thick as possible.

Plugs and sockets should be clean and well fitted with, preferably, soldered joints. Crimped joints should be made with the correct tool or reinforced with solder. If your soldering is at all suspect, get a more capable friend to do the job for you. Poor soldering can make the difference between success and failure.

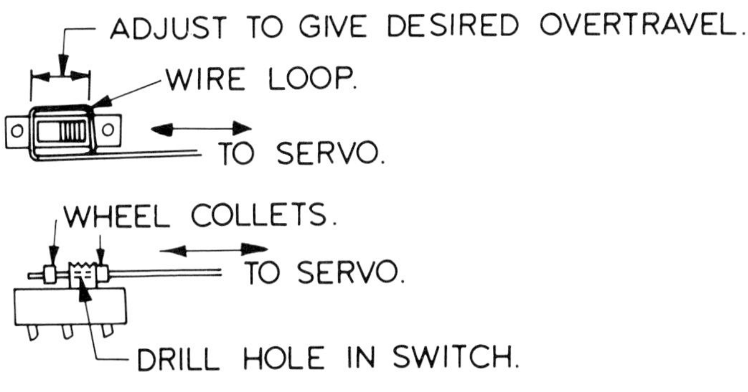

FIG. 47 Methods of operating slide switch of motor on/off.

FIG. 48 Suppressing relay contacts.

Chapter Eight

Installation

MOTORS

One very common method of installing the motor is to house it in a close-fitting balsa or cardboard tube built into the front of the model (see fig. 49).

The fit of the motor can be adjusted by wrapping it with masking tape. If the front of the motor coincides with the front of the fuselage, a good supply of cooling air will flow through the motor.

With GRP fuselages, the motor can conveniently be mounted to a plywood bulkhead as shown in fig. 50. Ensure that cooling holes are cut in the bulkhead so as to coincide with the holes in the front of the motor.

Where a more streamlined nose is preferred, it is possible to remove the front of a nylon spinner and, by drilling holes in the rear part, to arrange a supply of air through the spinner (see fig. 51)

Note that electric motors are virtually vibration free and it is unnecessary to resort to the positive attachment methods which would be required by an I.C. engine of similar power.

Photo 36. The motor on this scale subject is effectively friction-mounted between shaped balsa blocks, a simple system developed for electric round-the-pole flying.

FIG. 49 Simplest method of fitting motor.

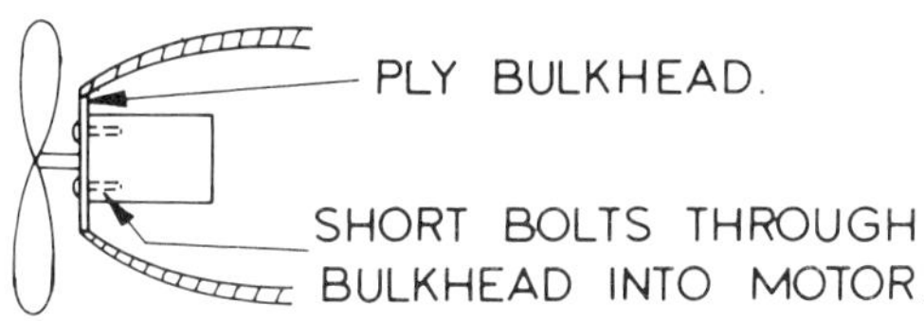

FIG. 50 Fitting motor to GRP fuselage.

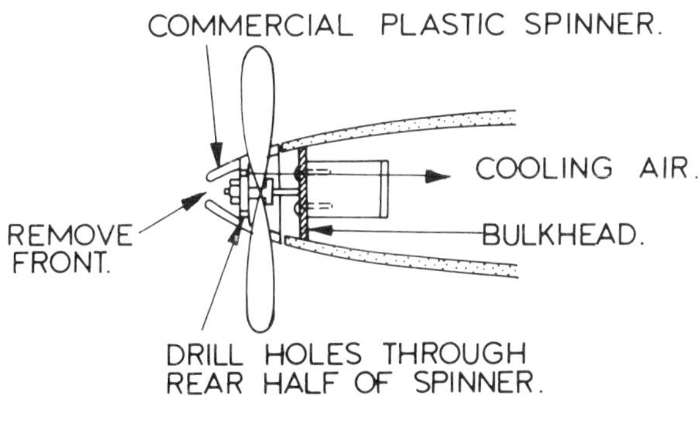

FIG. 51 More streamlined version.

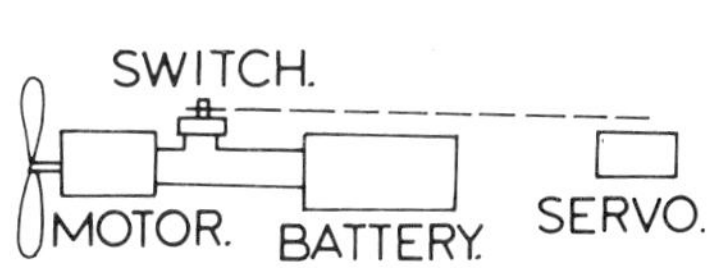

FIG. 52 Using conventional throttle linkage to on/off switch.

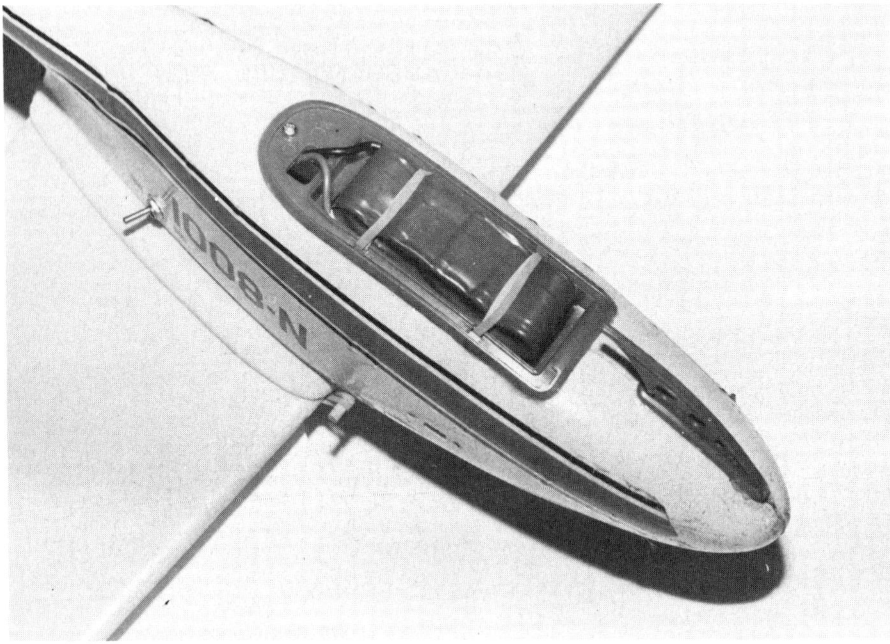

Photo 37. A wider view of the battery mounting on the Nitto 'Kitty', a practical system which allows survival in modest crashes and which is being followed in other models. Cool, too!

Do not assume, however, that the lack of noise and vibration means that electric motors are not dangerous. They are, and I still have the scars to prove it!

BATTERIES

Here the main requirement is to ensure that the battery will not cause damage to the airframe in the event of a heavy landing. One way of preventing this is to mount the battery outside the fuselage, as already mentioned.

If mounted inside, the battery should be wrapped in foam rubber and mounted against some substantial bulkhead such as a firewall or undercarriage mounting.

The position of the battery is, to some extent, dependent on the C of G position but is should be as close to the motor as possible, as previously discussed.

Ideally, the motor on/off switch should be positioned between the motor and the battery to reduce wiring. Where the switch is servo operated this can be done by means of a standard wire-in-tube type throttle linkage (see fig. 52).

A supply of cooling air to the battery should be arranged, remembering that every air *inlet* requires at least the same area of air *outlet.*

This is facilitated by the fact that it is unnecessary to seal off the nose area as with I.C. models.

Remember! No fuel proofing is required!

Chapter Nine

'Almost Ready to Fly' Models

ACOMS 'CESSNA'

This aircraft is almost certainly intended to appeal to the 'toy' market, yet it is quite a challenge to fly really well and should not really be regarded as a beginners' model.

All airframe components are moulded in polystyrene foam with a comprehensive set of vinyl 'decals' for decoration. Assembly consists of joining the wing halves with epoxy, gluing the tailplane, fin and undercarriage support into position, and bolting on the nosewheel and motor/gearbox mounting.

An RS380 type motor drives a 9in diameter propeller through reduction gearing and produces a reasonable rate of climb from the 42in span model which, with *Acoms* 2 channel radio weights 2lb (910gm).

There appears to be a considerable variation in the flight times reported for this model from various sources. The writer's example would, initially, only record flights of around 2½ minutes,

Photo 38. The Acoms Cessna is a very attractive model. Motor unit etc. are available separately for application to other models and an extra is a relay-operated (third channel) on/off switch.

Photo 39. The 61½ in Sirocco *from Acoms is simple to assemble and fly. Tip winglets are optional but advisable. Note single-wheel undercarriage.*

Photo 40. The Nitto Kitty *is an attractive little model and at 20-23 oz extremely light for its 50 in. span. Definitely not for rough-air flying.*

despite changing the motor and trying three different battery packs (6 cell, 600 mAH).

Replacing a broken propeller produced as instant improvement to around 4 minutes. while charging with the aid of a DVM as described in chapter 6 produced occasional flights of up to 6 minutes.

ACOMS SIROCCO

As a follow-up to the Cessna, the Acoms company launched (in mid 1983 in UK) a "kit" (in fact ARTF) of a free-lance powered sailplane type of model, based on fairly typical European motor-glider layout. At 1560mm span and a design weight of 1 kilogram with two-function radio, the standard RS380 geared to a 9×4 in. plastic prop at 3:1 ratio produces a pleasing climb and flights of 6-7 minutes in still air conditions, plus the ability to loop.

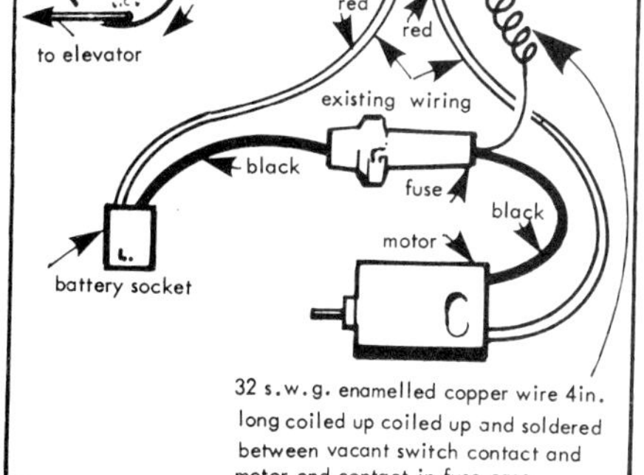

FIG 53 Switching by full elevator travel.

Construction is again all polystyrene foam with injection-moulded motor cradle and stressed components. The externally clipped 6-cell 600 MAH system is also employed, as on the Cessna. Flying characteristics are good, with things happening slowly enough to give a beginner time to think, but the model is strictly for light breeze at the most; it is at its best in near-calm thermally conditions when long flights should be possible. A successful motor on/off switch-operated by a quick pulse of full up or down elevator was devised by Peter Holland (see Fig 53) to incorporate dynamic braking (to stop the prop – a stationary prop has less drag than one windmilling – and allow flights to alternate between power and gliding spells.

NITTO 'KITTY'

Very similar in concept to the *Acoms* model, the 'Kitty' is quite a different aircraft, being a 50in span powered glider with a pusher propeller mounted above the wing.

Once again the construction is all of moulded foam and assembly follows very similar lines. The motor is an RS280 type (smallest yet) and drives a 7in diameter propeller through similar reduction gearing. Batteries are, like the 'Cessna', slung under the fuselage by rubber bands. Five cells of 250 or 450 mAH size (not supplied) can be accommodated, giving flights of around 2 and 5 minutes respectively.

With miniature R/C gear, the 'Kitty' weighs 20oz (568gm) with 250 mAH cells and 23oz (653gm) with the 450 mAh size.

With a high aspect ratio wing, low dihedral and a small rudder, the 'Kitty' is difficult to fly in turbulence, but on calm evenings it is great fun.

Chapter Ten

Commercial and Individual examples

APEX 98

Available as a kit or ready to fly (less motor, batteries and radio) the Apex 98 is from Edmonds Model Products of High Wycombe. There are actually three *Apex* models, 98 in. span rudder/elevator, 74 in. rudder/elevator and 74 in. with ailerons. Motor switching is either by separate servo or off the elevator servo, a blip of full down switching the motor on and full up motor off.

The unusual mounting of the Mabuchi 550 on the fin provides maximum propeller efficiency and has proved amply strong with the negligible vibration of electric power. It is counterbalanced by stowing the 8 1·2ah nicads in the nose, ahead of the radio gear; wiring losses using 30/0.25 wire are in the order of .25v. Using an 8×4 in. Taipan prop, power run is around 5 mins. on full charge, taking the model to a good height and allowing flights of 10-12 minutes without thermal assistance. Total weight of the *Apex* 98 with electric installation is 69 ozs. (1952 gms.).

Conversion kits are available for the three *Apex* models and the 100 in. *Aurora* from the same manufacturer and if these or motors, etc., are not available through a local model shop EMP will supply by mail.

Photo 41. The fin-mounted motor of the Apex 98 is unorthodox but the flight performance is excellent.

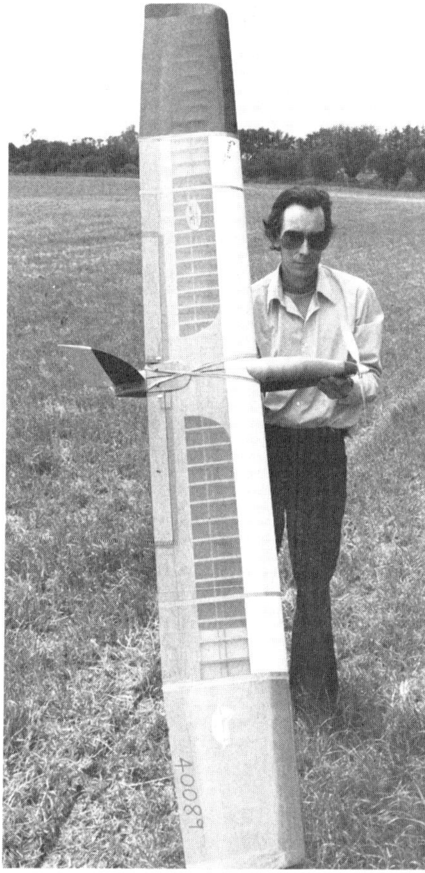

GOONEY BIRD

This "flying plank" type of model is a design by G. Johnson, available as a full-size plan from the M.A.P. range. Span is 99½ in. (2527mm) and the power unit is the Graupner 'Mosquito', 15 size, with 8 1·2 ah cells, which makes the model suitable for calmish weather thermal flights. As a slope soarer it can cope with winds of up to 15 m.p.h. Control is by rudder and elevators rather than elevons, and a characteristic of the model is an ability to make very tight, slow turns. Despite its appearance, it is a very forgiving model with a gentle stall and although the C.G. position is fairly critical, it is quite simple to trim. The power unit is used by the designer mainly to increase the rate of climb once centred in a weak thermal, or to climb to about 500ft. before switching off and gliding down to 300ft., when it is switched on again. Average flight is about 15 mins. and a useful tip is always to maintain a shallow climb with wide turns to avoid the motor labouring and consuming current faster.

Photo 42. Left, the Gooney Bird *can give many conventional gliders a run for their money in performance, stability and turning ability. Below, Photo 42. The Graupner 6v geared Jumbo 540 motor with elastic coupling in the nose of the Gooney Bird. Note the collet and rod on/off switch gear.*

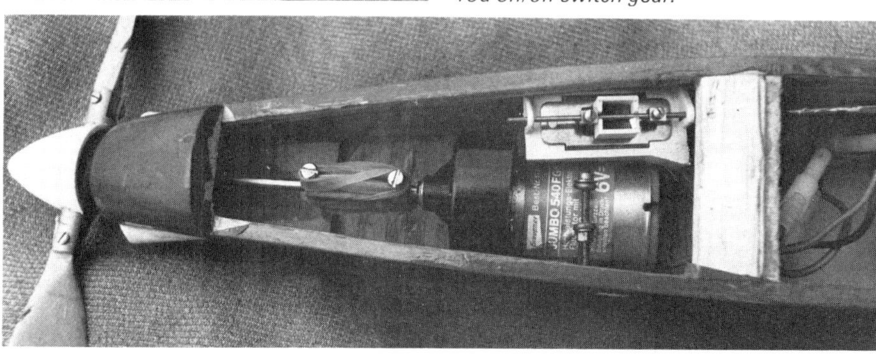

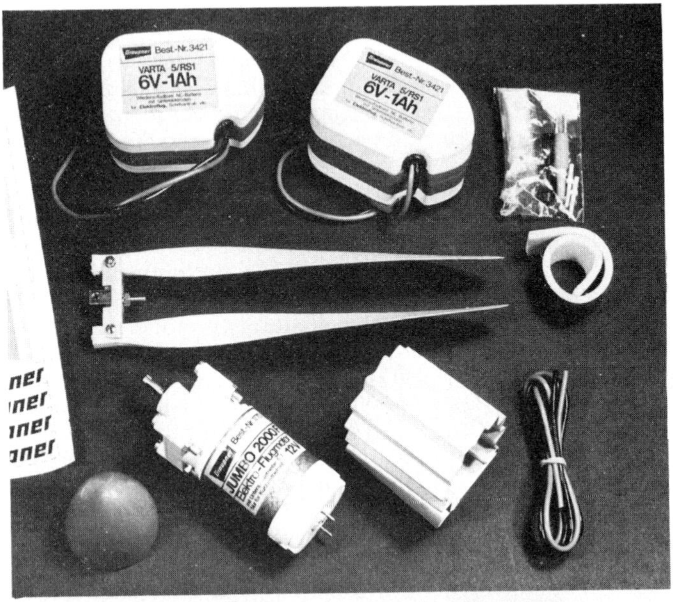

Photo 44. One of the early firms on the electric scene was Graupner, for whom the late Fred Militky worked. This flight pack includes a Jumbo 2000 (12v operation, but 6v and geared versions are available), air cooled motor mounting, folding prop and spinner and two 6v 1ah nicad packs.

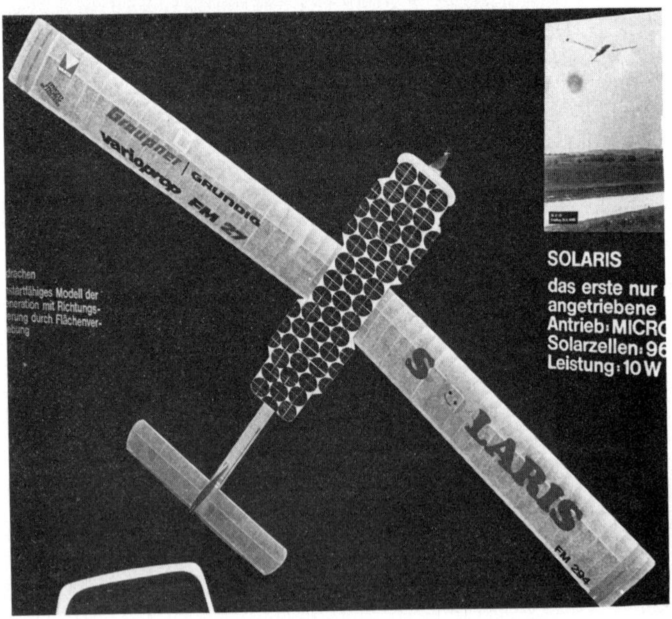

Photo 45. Solar-powered models have been developed by Graupner in Germany and, much larger, by the Boucher Brothers in the USA. However, cell costs are prohibitive at present – those for the model illustrated would cost over £330 (say $500).

JAMES WOT

This 72 in. (1830mm) two-function model looks almost like a large rubber-powered design. By Robin James, published in Radio Modeller, plans available from M.A.P. Ltd. Airframe weight is 12 ozs., flying weight with radio and Graupner Electroprop unit 2lb. 7ozs., average flight time 22 mins. including 7-8 mins power, enough to climb from ground level to 500ft. at least three times per charge. On/off is by latching microswitches operated by a blip of full rudder plus trim.

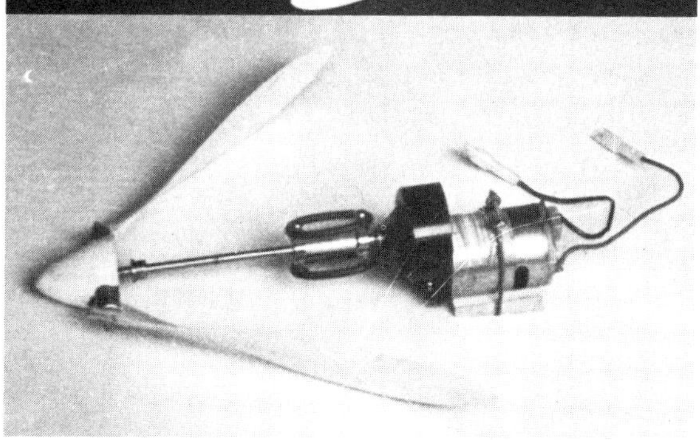

Photos 46, 47, 48. Clean lines and simple construction of James Wot *model are clear. Neat nose (with cooling intake) is aided by rubber-coupled extension shaft.*

SKYLARK EH-1 HELICOPTER

Two RS.540 motors and the customary 8 1.2 ah cells power this Ishimasa kit (rotor diameter 992mm., weight 1560 gms) but to learn to fly it is possible to use a 12v car battery connected by a long lead. Can be assembled in one evening, since the centre frame and tail rotor, etc. are factory built. Limitations are 3-4 mins. flight time between charges, unless an exterior battery and lead are used, and necessity for modern 4-function R/C with smallish servos.

ENGLISH ELECTRIC WREN

Scale model of a machine built for the 1923 Lympne light aeroplane trials. Original model with Olympus geared unit weighed 70 ozs. (span 92 in.), second with Graupner 'Mosquito' unit weighed 64 ozs. Will take off from smooth tarmac, climb to about 200 ft., cruise for 3-4 mins. and land after 7-8 mins. using 8 1.2 ah cells. Plan by J. Collins available from M.A.P.

AVRO LANCASTER

Remarkable 74 in. model by Peter Holland with four Cyclones switched either series or parallel from 8 2.5 ah cells, driving 6×4 in. nylon props. Total weight 6¼ lbs. Aileron and elevator control only, would take off and fly with fixed undercarriage or be hand-launched retracted, when over-travel of motor off switch would release undercarriage which dropped and self-locked.

EDELWEISS

This butterfly-tailed Robbe kit is 98½ in. span and weighs about 58 ozs. with a Robbe EF 76 II unit running off 7 cells and turning a 15 in. folding prop through 3.3:1 reduction. An average flight time of about 12 mins. can be expected. Latest Robbe information suggests that 8 cells are now recommended. A micro-switch provides off and electrical braking to fold the prop; running the batteries out completely is not advised.

GEIER

Robbe also produces this 60 in. flying wing in glider, i.c. engine or electric form, using an EF 76 II direct driving an 8×4 in. pusher propeller and 8×1.2 ah. Very stable and fast at 56 ozs. (glider version illustrated).

The Edelweiss *model up and away (above) and, below, the RCM test model being readied for a flight.*

MAGICFLY

A 48 in. aerobatic model from M.F.A., this model uses an M.F.A. Olympus unit. (a Hummingbird 15 motor with belt reduction of 2.3:1) swinging an 11×6 in. nylon prop with the customary 8×1.2 cells. Aileron and elevator controls are used and loops, rolls, etc. can be carried out, from level flight at the beginning but requiring a slight dive towards the end of the power run.

Two versions of Magicfly *are offered, with either foam or built-up wings; latter lighter and flight therefore improved. Note ply wingtip fences.*

ISLANDER

Two Hummingbird 15 motors and 8×1.2 cells power this 50 in. 64 oz. stand-off scale model from M.F.A. Uses ailerons and elevators – ailerons are essential, since an experimental model flying on rudder would not return to level flight when power fell off. Will take off from a smooth runway after a longish run. M.F.A. supply a charger with a recommended 30 min. charge time from a 12v battery.

ROBBE PARAT

Another Robbe model for the EF 76II, direct drive with a 7×4 in. prop and 7 or 8×1.2 cells. Batteries are stowed in a hatch beneath the fuselage. Span 50 in., weight around 48 oz., average still air flight time on test 8 mins.

COTSWOLD ELECTRIC TRACER

Mentioned earlier, the *Tracer* is a 37 in. Club 20 racer for which the Weston UK unit was originally produced. The manufacturers will supply kits of the race-proven model to order.

The Hummingbird *is a simple but neat little model using rudder and elevator control. A motor on/off switch operated by full rudder or elevator is a worthwhile inclusion.*

HUMMINGBIRD

This M.F.A. model was one of the earliest British kits capable of aerobatics, which tend to be crisper at the beginning of a flight when full power is available. At 44 in. span, a flying weight of 35-39 ozs. is achieved with a Hummingbird 15 and 8×600 ma/h cells, spinning a 6×3in. prop and giving flights of 8 – 10 mins. if flown gently. Sharp turns, climbs and aerobatics load the propeller more and use up the power quicker.

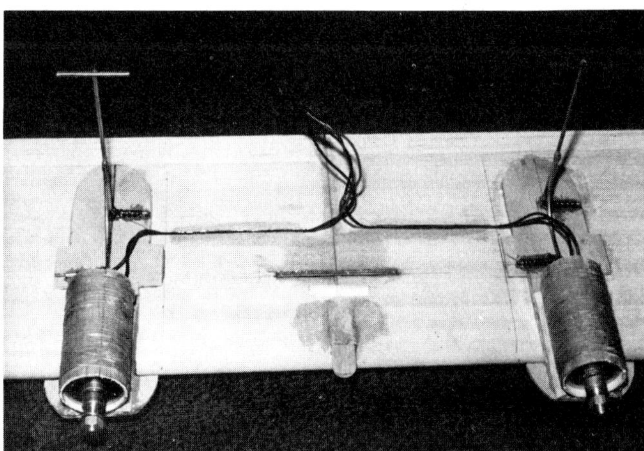

M.F.A. method of motor mounting (here on an Islander*) is to roll the motor tightly in 1/16 in. balsa sheet and bind with thread or tape, when it can be glued in place. Motor can be slid out. Avoid prolonged static running for fear of overheating.*

Photo 68. Light construction and a skid instead of an undercarriage are evident in this photo of Voltaire.

VOLTAIRE

One of the most popular plans, this model uses a Cyclone 15, 7×4 prop and 8×1.2 and weighs only 40 ozs. for 46 in. span. Very fast and capable of all aileron/elevator aerobatics, even one vertical roll or consecutive loops. Designed by J. Slade, plan available from M.A.P.

CHOP-STIK

An experimental 72 in. pusher model by Peter Holland with Cyclone unit and a 7×4 pusher prop, weight 45 ozs. Square wood boom changed to 3/8 in. alumin. tube to obviate whip. Inboard ailerons/elevator provided enough control for mild aerobatics

Photo 69. The intention with Chop-Stik *was to remove the 9 in. outboard wing panels for comparative tests, but a wing folding in an Immelman ended the programme.*

Photos 70-73 show the Hi-Fly, *special 12v charger, an Electroprop 2000 unit and a mixed adaptation using an .049, the electric motors being switched in as required when the .049 has run through its fuel. Very long flights possible!*

HI-FLY

Twin-pusher Graupner kit designed by the late Fred Militky using two Electroprop 2000 FD motors and two packs of 10 1.0 ah nicads giving a flying weight of 67 ozs. Folding 14 in. props are driven through 5:1 gearing and the motor mounts are simply (but adequately) stuck to the wing trailing edge by double-sided servo pads. Span is 91 ins. RCM test model was flown on one set of batteries (12v, .5 ah) which gave a power run in excess of 5 mins. and a checked initial rate of climb of 230 f.p.m. in winter conditions.

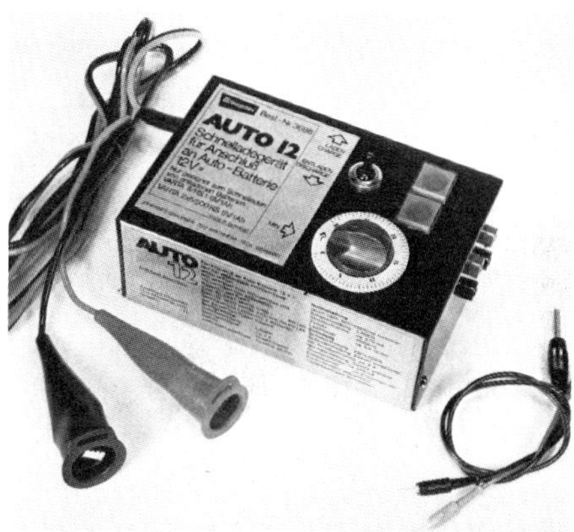

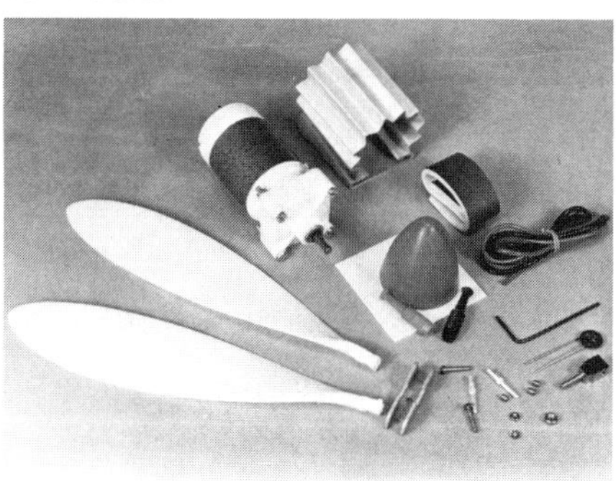

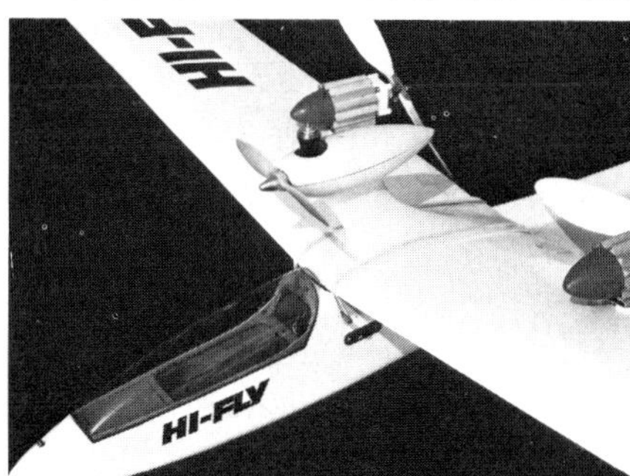

LEISURE PLAYBOY

Leisure Electronics is an American company which introduced a new range of motors, etc. and an annual competition in 1982. The "Playboy" is a 67½ in. version of a very popular 1940s pylon free-flight design. A test model built from a standard kit using a Leisure geared motor (3:1) driving an 11×7 prop and a miniature R/C installation weighed only 36 ozs. and provided an initial rate of climb of over 600 f.p.m. on six 1.2 nicads. The LT50 motor is available with different armature windings for direct or geared drive, with ball bearings on the shaft; the gearbox employs Delrin and brass gears also running on ball-raced shafts. A new, small aerobatic model under development as this book is prepared weighs 34 ozs., will give 10 min. flights and is capable of the full AMA Class A routine. American trend is apparently towards small aerobatic/small scale with direct drive and "old timer"/powered sailplanes with geared drive.

CORKY

This is a simple and attractive model from Galaxy Models, available as a glider kit (with built-up or foam wing option) and a conversion kit with an RS540 motor and all necessary parts. Span is 51in.

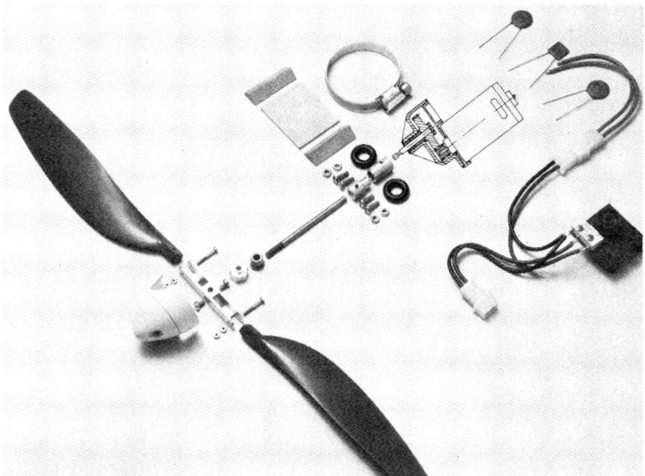

Robbe electric flight accessory set. Switch is on/off with dynamic braking at "off". Note use of Jubilee clip in mounting and rubber rings for shock-absorbing drive. Capacitors or chokes fit in each motor lead and between brushes for spark suppression.

REDUCTION

In general, it is more efficient to move a large amount of air slowly than a small amount quickly. Reduction of speed between motor and propeller allows the motor to run at a more efficient speed (i.e. more power for the same consumption, or the same power for longer, or the same power from lighter batteries) and improves airscrew efficiency. The choice is between gears or belt reduction; the latter usually requires less engineering skill to avoid losses, although the photos show a commercial (Hummingbird) belt unit and a home-made gearbox.

ACCESSORIES

Several manufacturers provide accessories but unless catalogues can be studied exactly what can be obtained may be difficult to establish. As an example, the Robbe range includes, in addition to aircraft kits and the Elt-Max motors (in direct drive or geared drive), a series of switches which mount on most rotary servos, chargers for mains and car battery input, two and three-blade folding props, suppressors, an extension shaft with bearing, rubber coupler, and spinner, wiring harness, relay modules and general items of use to electric fliers.

CONTINENTAL EXAMPLES

European contests for electric flight are well established and these pictures give a glimpse of the large and sophisticated models used.

About to become the first winner (1980) of the prestigious annual Militky Cup, Hans Schumacher's all-g.r.p. model is ready for launch. Hans, from Munich, designed the original Graupner Bellaphon R/C equipment, incidentally.

A French model by Roland Stuck, 4.2m span, 4.5kg, with a Geist 124 motor and 16 1.2 Sanyo cells. This range of motors, built by Fritz Geist in Germany, is little-known in Britain or U.S.A. The 124 can draw up to 25 amps on 10-16v, swinging a 16×6 prop at 4500 r.p.m. or so.

Roland Hersperger of Switzerland believes in size: this model is 7.3m (24ft.) span, weighs 9.5kg. (21 lbs.) and flies on a Keller 100 motor. At least the wings need not be stressed for bungee launches!

Summary

(For the non-technical)

There are still fields to explore and experiments to try. Dieter Grupe of Berlin won the 1983 Militky Trophy with this remarkable peripheral propeller, folding to create a super-clean machine.

IN SUMMARY

From the examples given in this book, it is clear that while it is entirely possible to have a racer or aerobatic model, a majority of kits and individual designs are basically powered soarers, offering flights of 12 mins. or so in calm air and twice as long or more when there is lift about. In this connection, incidentally, a model caught in an area of strong sink should be flown out of it as soon as possible, as the 200-250 ft. per minute climb rate of a reasonable model is not really able to cope with rapidly sinking air.

The average model size appears to fall in the 72 – 96in. span range, with an average weight of close to 4lbs. Most common motor for direct drive is the Mabuchi RS.540/550 series, but a gear reduction of 3:1 or thereabouts, allowing a much larger (preferably folding) airscrew, offers advantages for a little extra cost. Eight cells (9.6v nominal, 1.2ah) is the norm. Some form of on/off switch will prolong flights, extend the life of the cells, and offer a safety margin on landing approaches. If it shorts the motor brushes on "off", giving dynamic braking, positive propeller folding or reduced drag from a conventional airscrew will result. Light switches can be operated from full travel on one or other of the flying controls, avoiding the need for an additional servo, while if an electronic switch is used (see appendix) it is possible to dispense with separate radio batteries, saving an appreciable amount of weight.

Light weight must be the aim throughout. In a crash the batteries will tend to wreck the fuselage even if several ounces of reinforcement are included; mounting them in an easily-repaired bay underneath is worth considering. The absence of fuel spill, exhaust residue and vibration means a saving in weight in the covering and should assure long life. No fuel-proofing is an attraction to scale modellers, who can use an elaborate scheme without fear of it being ruined, but at present it seems advisable to stick to prototypes of the power soarer or ultra-light type. Further development will bring wider scope in the not-too-distant future, with more available power and/or longer flight times.

A common experience is improvement in motor/battery output after the first two or three flights. There is usually some surprise initially at the fast flat glide which requires a long unobstructed approach for landing, but once used to this flying should be possible on sites otherwise precluded by noise nuisance.

Appendix 1

An on/off in-flight switch and R/C battery eliminator

There are many ways of switching an electric motor on and off from one channel of a modern digital R/C outfit. The simplest method is to use a servo to move a slide or toggle switch. However, it seems a little bit pointless to use a motor to switch on a motor!

It is of course possible to switch on a motor directly from an R/C channel by means of power transistors. The problem here is that any transistor with give a voltage drop of approximately 0.6 volts and produce a fair amount of heat doing so. Add to this the fact that the more power you are switching, the bigger (and more expensive) this transistor must be. For electric flight purposes we wish to keep losses to a minimum which means that the humble relay has much to offer. A relay, in theory, has no losses, though after some use the contacts will become dirty and produce a voltage drop.

When this reaches the point where there is a noticeable loss of performance, it is easier to replace a couple of pounds' worth of relay than to delve into the electronics and replace several pounds' worth of power transistor. The relay can also be easily mounted between the battery and motor, thus keeping connections as short as possible.

First attempts were made by using a commercial servo IC which could operate a relay directly without any intermediary transistors. While this was successful, it was intended that the radio should be operated from the drive battery and this required a means of switching off the motor automatically before the voltage dropped to a level which would be dangerous to the radio. By adding a transistor between the servo IC and the relay it was possible to arrange for this to be operated by another IC such that the servo IC was disabled.

THEORY OF OPERATION

Referring to Fig. 1 the NE 544 servo amplifier IC (IC.3) is arranged to switch the relay in and out by means of transistor T2. Only a minimum of external components are needed to accomplish this with the switching point being determined by R9 and R10 which replace the normal servo pot. The values given (2.2k and 3.3k) ensure that the relay switching point will occur at 1.5ms which falls within the stick range of most R/C outfits available (see later). If you wish to make this point adjustable, simply replace these two resistors with a 5k preset potentiometer

The actual output from IC3 to T2 can be either from Pin 9 or Pin 13, this will depend on the throttle 'sense' of your particular R/C system. T2 can also be switched by IC2 which is a latching, voltage-sensitive switch. The point at which this operates is decided by the 100k preset potentiometer VR1. Once triggered, it can only be re-set by switching the supply off and on, which allows T1 to put a short positive pulse on to Pin 3 which resets the latch.

IC1 is a voltage regulator IC which supplies 5 volts for the R/C equipment. Note that the normal receiver battery is not required.

Diode D1 bypasses the back EMF from the relay coil and protects transistor T2 from possible damage.

The contacts of the relay specified are rated at 5 amps at 30v DC for 10,000,000 operations. Paralleling them gives 10amp rating

Several options are shown. A fuse is not really necessary since any dramatic load on the system should trigger IC2 and switch off the motor. The switch could be dispensed with by making the battery plug easily accessible.

On/Off In-flight Switch

FIG 1 Theoretical diagram.

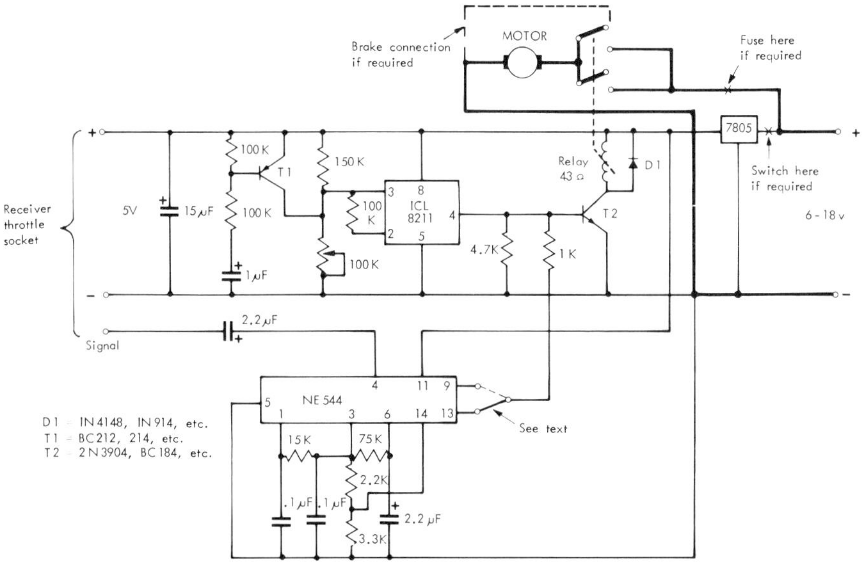

Below: first prototype was installed in Keller powered 'Tracer' using 12-1.2ah cells. Lower layer of 8 cells in two rows of four, with switch and relay mounted above front four cells and followed by two rows of two cells as shown.

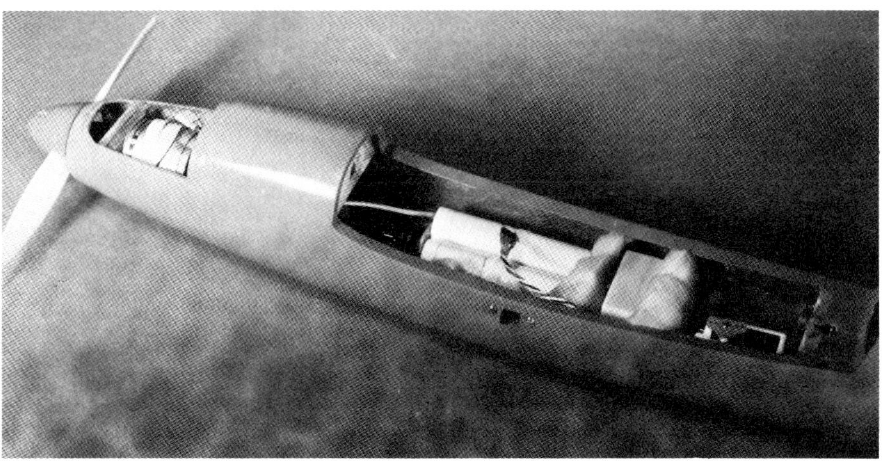

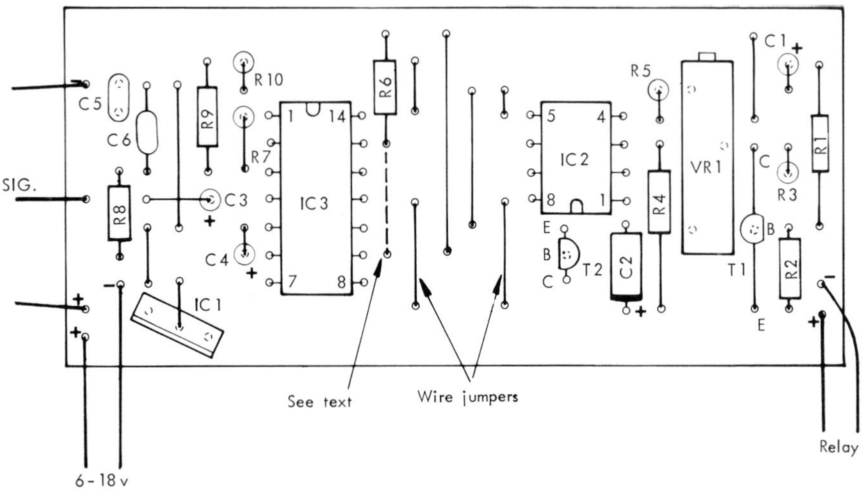

FIG 2 Component placement (N.B. copper lands on other side)

Some motors freewheel on the glide and if you are using an easily breakable prop or feel that the motor shaft may get bent on landing then the brake connection can be used.

CONSTRUCTION

1. Prepare the Veroboard by making the breaks in the copper lands as shown (a ⅛in.drill works nicely).
2. Install the ICs first as these help to position the various wire jumpers.
3. Add the jumpers next, then the resistors followed by the capacitors and, finally, the transistors.
4. You can solder as you go, or install everything and solder in one operation, it's up to you.
5. Install the seven wires using a different colour for each so they don't become confused.
6. Note that Fig. 2 and Fig. 3 show top and bottom sides of the Veroboard and are mirror images of each other.
7. Wire everything up as shown in Fig. 4 but don't connect the motor up for the moment.
8. Thoroughly clean the copper side of the Veroboard with cellulose thinners and an old toothbrush. Check for solder bridges.

Parts List

R1	100k	
R2	100k	
R3	100k	
R4	150k	
R5	4.7k	¼ watt
R6	1k	
R7	15k	
R8	75k	
R9	2.2k	
R10	3.3k	
C1	1μF	
C2	15μF	35v tantalum
C3	2.2μF	
C4	2.2μF	
C5	0.1μF	Disc ceramic
C6	0.1μF	
IC1	MC7805P	
IC2	ICL 8211 or RS 8211	
IC3	NE 544	
T1	BC 212, BC 214, BC 212L, BC 214L, etc	
T2	2N 3904, 2N 3906, BC 184, etc.	
D1	Any silicone diode (IN 4148, IN 914, etc)	
Relay	DPDT 6 volt type RS 349-642 53ohm coil (Radio spares); WE RL 113 43ohm coil (Watford Electronics)	
Veroboard	2.6in x 1.3in. x 0.1in. matrix	

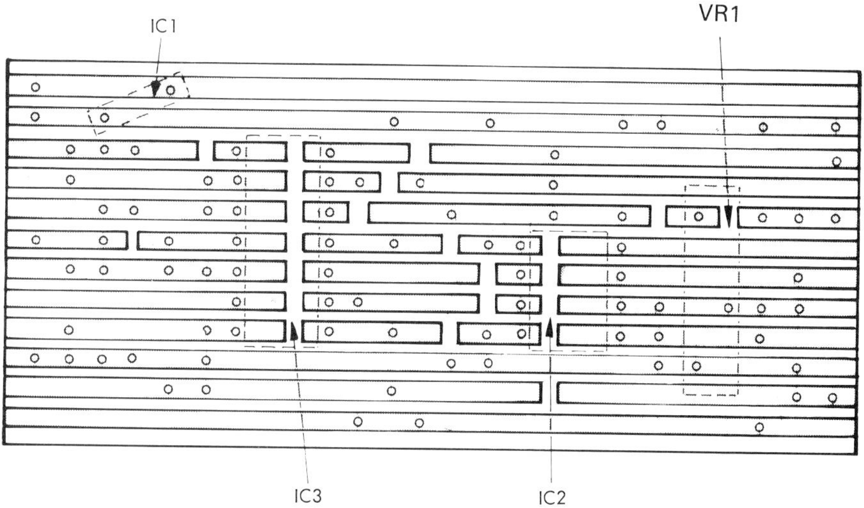

FIG 3 Veroboard layout showing breaks (approximately twice size).

TESTING AND SETTING UP

Plug in the drive battery, switch on and check for 5 volts at the receiver plug. If all is well, plug into the receiver and check that the receiver works by plugging a servo into another channel.

Adjust VR1 fully anti-clockwise and check that the throttle stick on your transmitter will operate the relay. Assuming that it does, check that the 'open' throttle position causes the relay to pull in. If it works in reverse (ie drops out on 'open' throttle) change R6 from Pin 9 on IC3 to Pin 13 or vice versa.

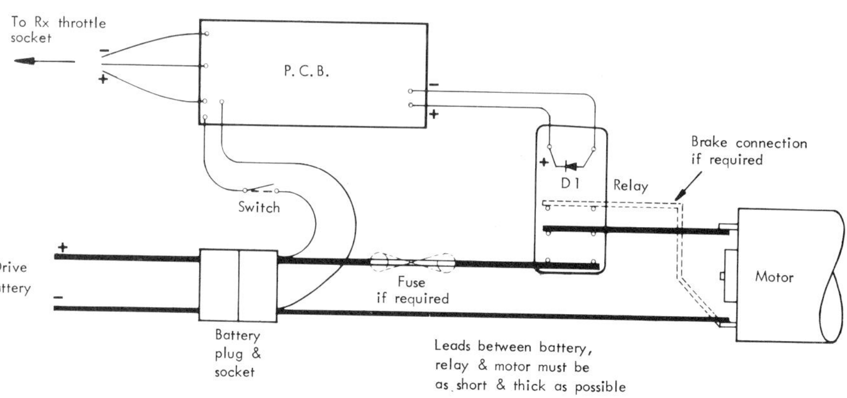

FIG 4 Connection diagram

Put the throttle stick to 'open' position so that the relay pulls in and slowly wind VR1 clockwise. At some point the relay should drop out. When it does turn VR1 slowly anti-clockwise until it pulls in again. Once again turn it clockwise and try to get it into a position roughly midway between the points where the relay pulls in and drops out.

Once you have established this position, turn VR1 clockwise until the relay drops out, then return it to the mid point already determined with the relay still out. Switching the supply on and off should now cause the relay to pull in. Assuming that all has worked so far, now connect up the motor and check that you can turn it on and off with your throttle stick. The motor should switch itself off when the drive battery shows signs of becoming discharged, at which point there should still be ample power to operate the radio equipment for some time.

FINDING FAULT

If you cannot get the relay to operate disconnect Pin 4 of IC2 from T2 by lifting one end of link 'A'. If it now works, the problem is in IC2 or associated components.

To check the operation of IC2, monitor the voltage at Pin 3. This should be above 1.2 volts. IC2 will trigger if it falls below 1.2 volts. If the voltage is higher than 1.2, momentarily short out T1 to check that this will reset IC2. If it does, the problem is in T1.

R9 and R10

As previously stated, these are chosen to give a switching point at approximately 1.5ms. If some other switching point is required, the appropriate values can be calculated as follows:

R10= (2 x time in ms)k;

R9= 5k – R10

Both values should then be corrected to the nearest standard resistor value *above* the calculated value.

OTHER USES

A forward and reverse system could be built by using two units arranged to operate at different pulse lengths and with one operating in the sense opposite (ie relay in with throttle stick in closed position). Only one unit would need to have IC1.

IC2 can be used on its own as a low voltage alarm system. Pin 4 will operate any LED directly.

All components shown from top (leads pointing downwards).

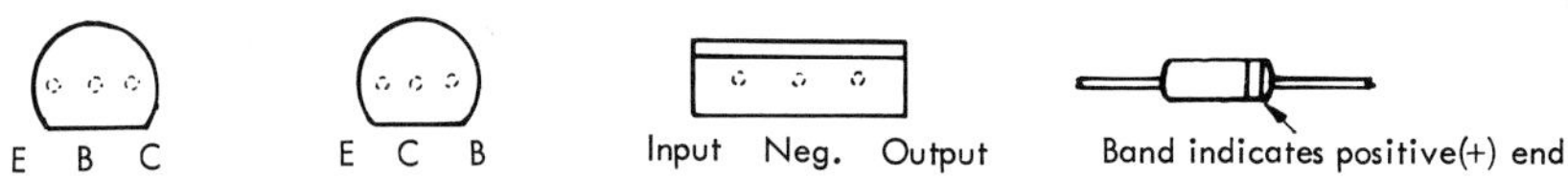

BC 214, BC 184, 2N 3904, etc. | BC 184L, BC 214L, etc. | IC1, 7805 | D1, 1N 4148, 1N 914, etc.

If you wish to retain the receiver battery supply and switch, simply omit IC1 and the connection to the drive battery.

For a twin motor, twin battery system, use two units with a 'Y' lead to the receiver but only connect *one* of the 5 volt positive lines to the receiver.

CONCLUSIONS

By omitting switches, fuses, receiver battery, etc., the lightest possible system can be built with the added virtue that the relay can be between the motor and batteries to reduce losses while the remaining components can be mounted anywhere in the model.

Appendix 2

Bibliography

Apart from "Silent Flight Special" published by M.A.P. Ltd. in 1977 (and long since out of print) British contributions to written matter on the subject of electric flight have mainly consisted of articles in Aeromodeller, Radio Control Models and Radio Modeller, and articles/chapters in Aeromodeller Annual and one other book. In addition there have been quite a number of model plans published in the above three magazines (most still available through M.A.P. Plans Service) plus reviews of kits and accessories. The following list is a selection of material of interest but is by no means fully comprehensive.

First all-electric takes the air	AM Sept 1957
Electric power	AM Mar 1960
Electric free flight	AM Annual 1960-1
Water-powered free flight	AM Annual 1962-3
Electric-powered radio control	AM Oct 1969
Mattel's ARTF Super Star	AM Dec 1972
Japanese model trade developments	AM Sept 1974
Electric motor suitability	RM Sept 1975 & Mar 1976
Electric power	AM Dec 1975
Quest 'Wisper'	AM Mar 1976
'Vampere' foam model	RM May 1976
Electric duration	RM Aug 1976
First all-electric fly-in	RM Dec 1976
YL Products electric flight system	AM Feb 1977
Current affairs	RM June 1978, Feb 1979 Sept 1979, Dec 1982

Encyclopedia of Model Aircraft (Octopus) (and republished for Marks & Spencer 1982)	1979
Electric Fly-ins	RM Dec.1979, Dec 1980, Feb 1981, Dec 1981
Electric Model Flying	RCM Apr/May 1981, Dec 1981
Acoms 'Cessna'	RCM Aug 1982, AM Sept 1982
Electrolight (with retracting motor)	AM Sept 1982
Acoms 'Sirocco'	RM Aug 1983
Electronic switch	RCM Aug 1983

Appendix 3

Glossary

Accumulator	An old name for a Lead/Acid wet cell.
A.C.	Alternating current.
Amp	Ampere. The unit of electrical current.
AH or Amp/Hour	Battery capacity expressed in terms of the current in Amps that it can supply for one hour.
Battery	A number of electrical cells connected in series.
Capacity.	The amount of electricity that a secondary cell can store.
Cell.	One unit of an electrical battery.
Charge	Means of replacing electricity in a discharged cell.
Charging.	Replacing electricity in a discharged cell.
Charger.	Device for charging
Current.	Amount of electricity flowing in a circuit, expressed in Amps.
D.C.	Direct Current.
Discharge	Removing electrical energy from a charged cell
D.V.M.	Digital Volt Meter.
E.M.F.	Electromotive force. Electrical energy.
Fuse.	A protective device to limit the current which can flow in a circuit.

G.R.P.	Glass Reinforced Plastic. Glassfibre.
Motor	A device for converting electrical energy into mechanical energy.
mA	Milliampere. One thousandth part of an Amp
Parallel.	A number of cells connected such that the voltage remains that of a single cell, but the capacity is multiplied by the number of cells.
Propeller.	A rotating wing which propels an aircraft through the air.
Rating.	The nominal capacity of a cell or the number of Watts which can be carried by given size of wire.
R/C.	Radio Control.
Reverse Charge	Passing electrical energy into a secondary cell with a polarity opposite to that for which the cell is designed.
R.F.	Radio Frequency.
R.F. Interference or R.F.I.	High frequency waves emitted by an electrical discharge which cause interference to radio transmissions.
Switch	A means of interrupting the flow of an electric current.
Suppression	Preventing the emission of R.F.I.
Series	A number of cells connected in such a way that the voltage is multiplied by the number of cells while the capacity remains that of one cell.
Volt.	The unit of electrical pressure which is present in a circuit.
Watt.	The unit of electrical energy. Equal to Volts x Amps